Sadaaki Miyamoto, Hidetomo Ichihashi, Katsuhiro Honda

Algorithms for Fuzzy Clustering

Studies in Fuzziness and Soft Computing, Volume 229

Editor-in-Chief

Prof. Janusz Kacprzyk
Systems Research Institute
Polish Academy of Sciences
ul. Newelska 6
01-447 Warsaw
Poland
E-mail: kacprzyk@ibspan.waw.pl

Further volumes of this series can be found on our homepage: springer.com

Vol. 214. Irina Georgescu
Fuzzy Choice Functions, 2007
ISBN 978-3-540-68997-3

Vol. 215. Paul P. Wang, Da Ruan, Etienne E. Kerre (Eds.)
Fuzzy Logic, 2007
ISBN 978-3-540-71257-2

Vol. 216. Rudolf Seising
The Fuzzification of Systems, 2007
ISBN 978-3-540-71794-2

Vol. 217. Masoud Nikravesh, Janusz Kacprzyk, Lofti A. Zadeh (Eds.)
Forging New Frontiers: Fuzzy Pioneers I, 2007
ISBN 978-3-540-73181-8

Vol. 218. Masoud Nikravesh, Janusz Kacprzyk, Lofti A. Zadeh (Eds.)
Forging New Frontiers: Fuzzy Pioneers II, 2007
ISBN 978-3-540-73184-9

Vol. 219. Roland R. Yager, Liping Liu (Eds.)
Classic Works of the Dempster-Shafer Theory of Belief Functions, 2007
ISBN 978-3-540-25381-5

Vol. 220. Humberto Bustince, Francisco Herrera, Javier Montero (Eds.)
Fuzzy Sets and Their Extensions: Representation, Aggregation and Models, 2007
ISBN 978-3-540-73722-3

Vol. 221. Gleb Beliakov, Tomasa Calvo, Ana Pradera
Aggregation Functions: A Guide for Practitioners, 2007
ISBN 978-3-540-73720-9

Vol. 222. James J. Buckley, Leonard J. Jowers
Monte Carlo Methods in Fuzzy Optimization, 2008
ISBN 978-3-540-76289-8

Vol. 223. Oscar Castillo, Patricia Melin
Type-2 Fuzzy Logic: Theory and Applications, 2008
ISBN 978-3-540-76283-6

Vol. 224. Rafael Bello, Rafael Falcón, Witold Pedrycz, Janusz Kacprzyk (Eds.)
Contributions to Fuzzy and Rough Sets Theories and Their Applications, 2008
ISBN 978-3-540-76972-9

Vol. 225. Terry D. Clark, Jennifer M. Larson, John N. Mordeson, Joshua D. Potter, Mark J. Wierman
Applying Fuzzy Mathematics to Formal Models in Comparative Politics, 2008
ISBN 978-3-540-77460-0

Vol. 226. Bhanu Prasad (Ed.)
Soft Computing Applications in Industry, 2008
ISBN 978-3-540-77464-8

Vol. 227. Eugene Roventa, Tiberiu Spircu
Management of Knowledge Imperfection in Building Intelligent Systems, 2008
ISBN 978-3-540-77462-4

Vol. 228. Adam Kasperski
Discrete Optimization with Interval Data, 2008
ISBN 978-3-540-78483-8

Vol. 229. Sadaaki Miyamoto, Hidetomo Ichihashi, Katsuhiro Honda
Algorithms for Fuzzy Clustering, 2008
ISBN 978-3-540-78736-5

Sadaaki Miyamoto,
Hidetomo Ichihashi, Katsuhiro Honda

Algorithms for Fuzzy Clustering

Methods in c-Means Clustering with Applications

Springer

Authors

Dr. Sadaaki Miyamoto
University of Tsukuba
Inst. Information Sciences
and Electronics
Ibaraki
305-8573 Japan
Email: miyamoto@risk.tsukuba.ac.jp

Dr. Katsuhiro Honda
Osaka Prefecture University
Graduate School of Engineering
1-1 Gakuen-cho
Sakai
Osaka, 599-8531 Japan

Dr. Hidetomo Ichihashi
Osaka Prefecture University
Graduate School of Engineering
1-1 Gakuen-cho
Sakai
Osaka, 599-8531 Japan

ISBN 978-3-540-78736-5 e-ISBN 978-3-540-78737-2

DOI 10.1007/978-3-540-78737-2

Studies in Fuzziness and Soft Computing ISSN 1434-9922

Library of Congress Control Number: 2008922722

Typeset & Cover Design: Scientific Publishing Services Pvt. Ltd., Chennai, India.

Printed in acid-free paper

9 8 7 6 5 4 3 2 1

springer.com

Preface

Recently many researchers are working on cluster analysis as a main tool for exploratory data analysis and data mining. A notable feature is that specialists in different fields of sciences are considering the tool of data clustering to be useful. A major reason is that clustering algorithms and software are flexible in the sense that different mathematical frameworks are employed in the algorithms and a user can select a suitable method according to his application. Moreover clustering algorithms have different outputs ranging from the old dendrograms of agglomerative clustering to more recent self-organizing maps. Thus, a researcher or user can choose an appropriate output suited to his purpose, which is another flexibility of the methods of clustering.

An old and still most popular method is the K-means which use K cluster centers. A group of data is gathered around a cluster center and thus forms a cluster. The main subject of this book is the fuzzy c-means proposed by Dunn and Bezdek and their variations including recent studies. A main reason why we concentrate on fuzzy c-means is that most methodology and application studies in fuzzy clustering use fuzzy c-means, and fuzzy c-means should be considered to be a major technique of clustering in general, regardless whether one is interested in fuzzy methods or not. Moreover recent advances in clustering techniques are rapid and we require a new textbook that includes recent algorithms. We should also note that several books have recently been published but the contents do not include some methods studied herein.

Unlike most studies in fuzzy c-means, what we emphasize in this book is a family of algorithms using entropy or entropy-regularized methods which are less known, but we consider the entropy-based method to be another useful method of fuzzy c-means. For this reason we call the method of fuzzy c-means by Dunn and Bezdek as the standard method to distinguish it from the entropy-based method. Throughout this book one of our intentions is to uncover theoretical and methodological differences between the standard method and the entropy-based method. We do note claim that the entropy-based method is better than the standard method, but we believe that the methods of fuzzy c-means become *complete* by adding the entropy-based method to the standard one by Dunn and Bezdek, since we can observe natures of the both methods more deeply by contrasting these two methods.

Readers will observe that the entropy-based method is similar to the statistical model of Gaussian mixture distribution since both of them are using the error functions, while the standard method is very different from a statistical model. For this reason the standard method is purely fuzzy while the entropy-based method connects a statistical model and a fuzzy model.

The whole text is divided into two parts: The first part that consists of Chapters 1∼5 is theoretical and discusses basic algorithms and variations. This part has been written by Sadaaki Miyamoto.

The second part is application-oriented. Chapter 6 which has been written by Hidetomo Ichihashi studies classifier design; Katsuhiro Honda has written Chapters 7∼9 where clustering algorithms are applied to a variety of methods in multivariate analysis.

The authors are grateful to Prof. Janusz Kacprzyk, the editor, for his encouragement to contribute this volume to this series and helpful suggestions throughout the publication process. We also thank Dr. Mika Sato-Ilic and Dr. Yasunori Endo for their valuable comments to our works.

We finally note that studies related to this book have partly been supported by the Grant-in-Aid for Scientific Research, Japan Society for the Promotion of Science, No.16300065.

January 2008

Sadaaki Miyamoto
Hidetomo Ichihashi
Katsuhiro Honda

Contents

1 Introduction

The word of *a cluster* which implies a bunch of things of the same kind or a group of similar things is becoming popular now in a variety of scientific fields. This word has different technical meanings in different disciplines, but what we study in this book is *cluster analysis* or *data clustering* which is a branch in data analysis and implies a bundle of algorithms for *unsupervised classification*. For this reason we use the term of *cluster analysis*, *data clustering*, *unsupervised classification* exchangeably and we frequently call it simply *clustering*, as many researchers do.

Classification problems have been considered in both classical and Bayesian statistics [83, 30], and also in studies in neural networks [10]. A major part of studies has been devoted to supervised classification in which a number of classes of objects are given beforehand and an arbitrary observation should be allocated into one of the classes. In other words, a set of classification rules should be derived from a set of mathematical assumptions and the given classes. Unsupervised classification problems are also mentioned or considered in most textbooks at the same time (e.g., [83, 10, 30]). In an unsupervised classification problem, no predefined classes are given but data objects or individuals should form a number of groups so that distances between a pair of objects within a group should be relatively small and those between different groups should be relatively large.

Clustering techniques have long been studied and there are a number of books devoted to this subjects, e.g., [1, 35, 72, 80, 150] (we do not refer to books on fuzzy clustering and SOM(self-organizing map), as we will cite them later). These books classify different clustering techniques according to their own ideas, but we will first mention the two classical classes of hierarchical and nonhierarchical techniques discussed in Anderberg [1]. We have three reasons to take these two classes. First, the classification is simple since it has only two classes. Second, each of the two classes has a typical method: the agglomerative hierarchical method in the class of hierarchical clustering, and the method of K-means in the class of nonhierarchical clustering. Moreover each has its major origin. For the hierarchical clustering we can refer to the numerical taxonomy [149].

S. Miyamoto et al.: Algorithms for Fuzzy Clustering, STUDFUZZ 229, pp. 1–7, 2008.
springerlink.com

Although it is old, its influence continues until nowadays. For the nonhierarchical clustering, we can mention an old book by Duda and Hart [29] and a work by Ball and Hall [2] in which the old name of ISODATA is found, and another well-known work by MacQueen [95] where the concept of K-means has been proposed.

This book is devoted to the discussion of crisp and fuzzy c-means and related techniques; the latter is a fuzzy version of the K-means, while the original nonfuzzy technique is called the crisp c-means or hard c-means, as the number of clusters is denoted by c instead of K. Why we consider the c-means clustering here is justified by general understanding that a major part of researches has been done on and around this technique, while we do not discuss hierarchical clustering in detail, as hierarchical techniques have already been discussed by one of the authors [99], where how concepts in fuzzy sets are essential and useful in agglomerative hierarchical clustering have been shown.

Our primary motivation for this book is to show a less-known method of fuzzy c-means together with the well-known one: the former uses an entropy-based criterion while the latter employs the objective function by Dunn [31, 32] and Bezdek [5, 6]. As the both uses the alternate optimization with respect to the membership matrix and the cluster centers and moreover the constraint is the same for the both, the difference between the two methods is the objective functions.

We have a number of reasons why we discuss the entropy-based method [90, 91, 102, 68].

1. Methods using entropy functions have been rediscovered repeatedly in fuzzy clustering by different formulations.
2. This method is related to the general principle of maximum entropy [170] that has the potentiality of further development and various applications.
3. The method of entropy is closely related to statistical models such as the Gaussian mixture model [98] and the Gibbs distribution [134].
4. Comparison between the method by Dunn and Bezdek and the entropy-based algorithm more clearly reveals different features of the two methods.

We will observe the method of Dunn and Bezdek which we also call the standard method of fuzzy c-means is purely fuzzy, while the entropy-based method is more similar to statistical models.

Given a set of data, we have to use different techniques of clustering in general. In such a case knowledge on relations among different methods will be useful to predict what types of outputs will be available before actual application of an algorithm. For such a purpose theoretical studies are useful, since theoretical properties enable general prediction on results of clustering. For example, the Voronoi regions produced by crisp and fuzzy c-means clustering are useful to recognize natures of clusters which will be discussed later in this book. As another example, we will introduce fuzzy classification functions, or in other words, fuzzy classifiers, in order to observe differences between the standard and entropy-based methods of fuzzy c-means.

1.1 Fuzziness and Neural Networks in Clustering

Active researches in fuzzy sets [175] and neural networks, in particular, Kohonen's self-organizing maps [85] strongly stimulated pattern classification studies, and many new techniques of clustering have been developed. We now have a far longer list of methods in clustering and the list is still growing. No doubt this tendency is highly appreciated by methodologically-oriented researchers. However, we have a naïve question: Are methods using fuzziness and/or neural networks really useful in applications? In other words, Isn't it sufficient to consider traditional methods of K-means and statistical models alone? Two attitudes are taken by application scientists/engineers. If a convenient software of a method is available, they are ready to use the method. A typical example is MATLAB where the standard fuzzy c-means and the mountain clustering [173] are implemented, and S-PLUS where the FANNY [80] program can be used (the algorithm of FANNY is complicated and to develop the program is not an easy task). The other approach is to develop a domain-dependent technique by referring to traditional methods; good examples are BIRCH [180] and LINGO [125] by which clustering of large databases or that for information retrieval is studied. Such methods are, however, frequently based on the classical K-means with hierarchy formation and database structure consideration or use the traditional singular value decomposition [43].

New techniques will thus be really useful to the former class of users in applications by preparing a good quality software, while the implications and theoretical essences should be conveyed to the latter class of domain-oriented researchers.

Fuzzy clustering techniques are already found to be useful to the former, as some methods are implemented into well-known software packages. To the latter, we should further develop the theory of fuzzy clustering and uncover its relations to and differences from classical frameworks, and the same can be said on neural network techniques.

This book should be used by researchers and students who are interested in fuzzy clustering and neural network clustering; it should also be useful to more general readers in the latter methodological sense. For example, the entropy-based method relates a family of statistical models and fuzzy c-means by using the well-known fact that maximization of an entropy function with the constraint on its variance leads to the Gaussian probability density function. Another methodological example in this book is a recent technique in fuzzy c-means related to a new method in neural networks, that is, the use of kernel functions in support vector machines [163, 164, 14, 20].

The first part of this book thus shows less-known methods as well as standard algorithms, whereas the second part discusses the use of fuzzy c-means and related methods to multivariate analysis. Here the readers will find how techniques in neural networks and machine learning are employed to develop new methods in combination of clustering and multivariate analysis.

1.2 An Illustrative Example

Before formal considerations in subsequent chapters, we show a simple illustrative example in order to show a basic idea of how an agglomerative method and c-means clustering algorithm work. Readers who already have basic knowledge of cluster analysis may skip this section.

Example 1.1. Let us observe Figure 1.1 where five points are shown on a plane. The coordinates are

$$x_1 = (0, 1.5),\ x_2 = (0, 0),\ x_3 = (2, 0),\ x_4 = (3, 0),\ x_5 = (3, 1.5).$$

We classify these five points using two best known methods: the single link [35] in agglomerative hierarchical clustering, and the crisp c-means in the class of nonhierarchical algorithms.

The Single Link

We describe the clustering process informally here; the formal definition is given in Section 5.3.

An algorithm of agglomerative clustering starts from regarding each point as a cluster. That is, $G_i = \{x_i\}$, $i = 1, 2, \ldots, 5$. Then two clusters (points) of the closest distance are merged into one cluster. Since x_3 and x_4 has the minimum distance 1.0 among all distances for every pair of points, they are merged into $G'_3 = \{x_3, x_4\}$. After the merge, a distance between two clusters should be defined. For clusters of single points like $G_j = \{x_j\}$, $j = 1, 2$, the definition is trivial:

$$d(G_1, G_2) = \|x_1 - x_2\|,$$

where $d(G_1, G_2)$ is the distance between the clusters and $\|x_1 - x_2\|$ is the Euclidean distance between the two points. In contrast, the definition of $d(G'_3, G_k)$, $k = 1, 2, 5$ is nontrivial. Among possible choices of $d(G'_3, G_k)$ in Section 5.3, the

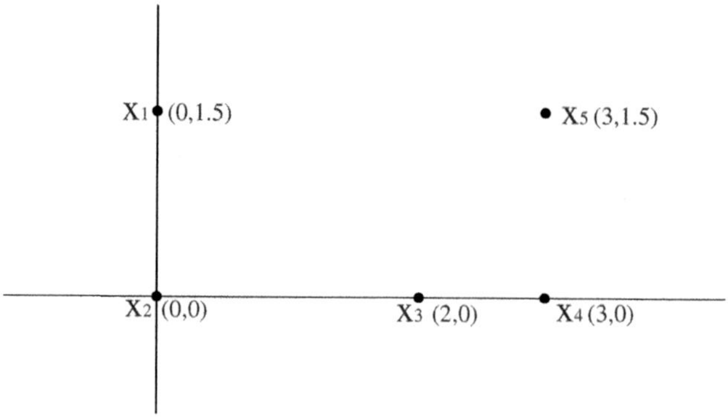

Fig. 1.1. An illustrative example of five points on the plane

single link defines the distance between the clusters to be the minimum distance between two points in the different clusters. Here,

$$d(G'_3, G_k) = \min\{\|x_3 - x_k\|, \|x_4 - x_k\|\}, \qquad k = 1, 2, 5.$$

In the next step we merge the two clusters of closest distance again. There are two choices: $d(G_1, G_2) = 1.5$ and $d(G'_3, G_5) = 1.5$. In such a case one of them is selected. We merge $G'_1 = G_1 \cup G_2 = \{x_1, x_2\}$ and then the distance between G'_1 and other clusters are calculated by the same rule of the minimum distance between two points in the respective clusters. In the next step we have $G''_3 = \{x_3, x_4, x_5\}$. Figure 1.2 shows the merging process up to this point, where the nested clusters G'_3 and G''_3 are shown. Finally, the two clusters G'_1 and G''_3 are merged at the level

$$d(G'_1, G''_3) = \min_{x \in G'_1, x' \in G''_3} \|x - x'\| = \|x_2 - x_3\| = 2.0$$

and we obtain the set of all points as one cluster.

To draw a figure like Fig. 1.2 is generally impossible, as points may be in a high-dimensional space. Hence the output of an algorithm of agglomerative hierarchical clustering has the form of a tree called a *dendrogram*.

The dendrogram of the merging process of this example is shown in Figure 1.3, where a leaf of the tree is each point. A merge of two clusters is shown as a branch and the vertical axis shows the level of the merge of two clusters. For example, the leftmost branch that shows x_1 and x_2 are merged is at the vertical level 1.5 which implies $d(G_1, G_2) = 1.5$.

In this way, an agglomerative clustering outputs a dendrogram that has detailed information on the merging process of nested clusters which are merged one by one. Hence it is useful when detailed knowledge of clusters for a relatively small number of objects is needed.

On the other hand, the dendrogram may be cumbersome when handling a large number of objects. Imagine a dendrogram of a thousand objects!

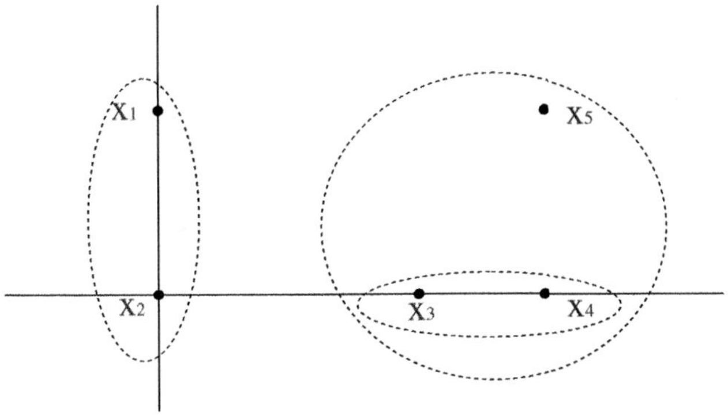

Fig. 1.2. Nested clusters of the five points by the single link method

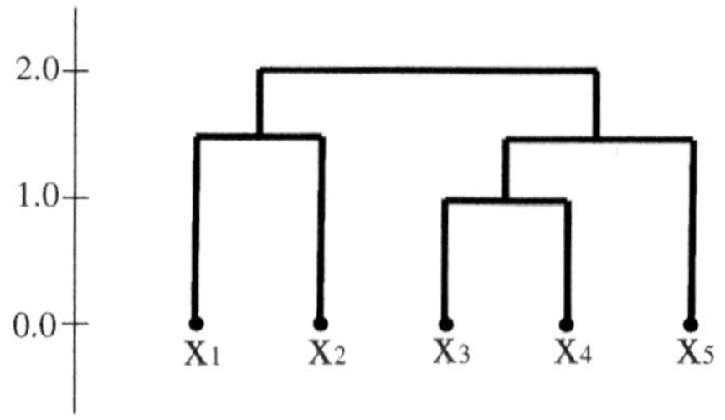

Fig. 1.3. The dendrogram of the clusters in Fig. 1.2

Crisp c-Means Clustering

In an application of many objects, the best-known and simple algorithm is the method of c-means. To distinguish from fuzzy c-means, we sometimes say it the crisp c-means or the hard c-means.

The parameter c is the number of clusters which should be given beforehand. In this example we set $c = 2$: we should have two clusters. A simple algorithm of crisp c-means starts from a partition of points which may be random or given by an *ad hoc* rule. Given the initial partition, the following two steps are repeated until convergence, that is, no change of cluster memberships.

I. Calculate the center of each cluster as the center of gravity which is also called the *centroid.*
II. Reallocate every point to the nearest cluster center.

Let us apply this algorithm to the example. We assume initial clusters are given by $G_1\{x_1, x_2, x_3\}$ and $G_2 = \{x_4, x_5\}$. Then the cluster centers v_1 and v_2 are calculated which is shown by circles in Figure 1.4. The five points are then

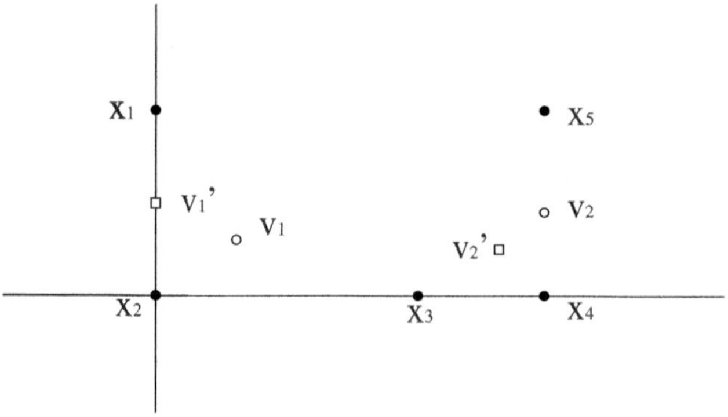

Fig. 1.4. Two cluster centers by the c-means algorithm with $c = 2$

reallocated. x_1 and x_2 are in the cluster of the center v_1, while x_3, x_4, and x_5 should be allocated to the cluster of the center v_2, that is, x_3 changes its membership. The new cluster centers are calculated again; they are shown by squares with the labels v_1' and v_2'. Now, x_1 and x_2 are to the cluster center v_1' whereas the other three are to v_2', i.e., we have no change of cluster memberships and the algorithm is terminated.

This algorithm of c-means is very simple and has many variations and generalizations, as we will see in this book.

2 Basic Methods for c-Means Clustering

Perhaps the best-known method for nonhierarchical cluster analysis is the method of K-means [95] which also is called the crisp c-means in this book.

The reason why the c-means clustering has so frequently been cited and employed is its usefulness as well as the potentiality of this method, and the latter is emphasized in this chapter. That is, the idea of the c-means clustering has the potentiality of producing various other methods for the same or similar purpose of classifying data set without an external criterion which is called unsupervised classification or more simply data clustering.

Thus clustering is a technique to generate groups of subsets of data in which a group called *cluster* is *dense* in the sense that a distance within a group is small, whereas a distance between clusters is *sparse* in that two objects from different clusters are distant. This vague statement is made clear in the formulation of a method.

On the other hand, we have the fundamental idea to use fuzzy sets to clustering. Why the idea of fuzzy clustering is employed is the same as above: not only its usefulness but also its potentiality to produce various other algorithms, and we emphasizes the latter.

The fuzzy approach to clustering is capable of producing many methods and algorithms although fuzzy system for the present purpose does not have profound mathematical structure in particular. The reason that the fuzzy approach has such capability is its inherent feature of linking/connecting different methodologies including statistical models, machine learning, and various other heuristics.

Therefore we must describe not only methods of fuzzy clustering but also connections to other methods.

In this chapter we first describe several basic methods of c-means clustering which later will be generalized, modified, or enlarged to a larger family of related methods.

2.1 A Note on Terminology

Before describing methods of clustering, we briefly review terminology used here.

S. Miyamoto et al.: Algorithms for Fuzzy Clustering, STUDFUZZ 229, pp. 9–42, 2008.
springerlink.com

First, a set of *objects* or *individuals* to be clustered is given. An object set is denoted by $X = \{x_1, \dots, x_N\}$ in which x_k, $(k = 1, 2, \dots, N)$ is an object. With a few exceptions, $x_1, \dots, x_N$ are vectors of real p-dimensional space $\boldsymbol{R}^p$. A generic element $x \in \boldsymbol{R}^p$ is the vector with real components $x^1, \dots, x^p$; we write $x = (x^1, \dots, x^p) \in \boldsymbol{R}^p$.

Two basic concepts used for clustering are *dissimilarity* and *cluster center*.

As noted before, clustering of data is done by evaluating nearness of data. This means that objects are placed in a topological space, and the nearness is measured by using a dissimilarity between two objects to be clustered. A dissimilarity between an arbitrary pair $x, x' \in X$ is denoted by $D(x, x')$ which takes a real value.

This quantity is symmetric with respect to the two arguments:

$$D(x, x') = D(x', x), \qquad \forall x, x' \in X. \tag{2.1}$$

Since a dissimilarity measure quantifies nearness between two objects, a small value of $D(x, x')$ means x and x' are near, while a large value of $D(x, x')$ means x and x' are distant. In particular, we assume x is nearest to x itself:

$$D(x, x) = \min_{x' \in X} D(x, x'). \tag{2.2}$$

In relation to a dissimilarity measure, we also have a concept of *metric*, which is standard in many mathematical literature (e.g. [94]). Notice that a metric $m(x, y)$ defined on a space S satisfies the following three axioms:

(i) $m(x, y) \geq 0$ and $m(x, y) = 0 \iff x = y$;
(ii) $m(x, y) = m(y, x)$;
(iii) [triangular inequality] $m(x, y) \leq m(x, z) + m(z, y)$.

Remark that a metric can be used as a dissimilarity, while a dissimilarity need not be a metric, that is, the axiom (iii) is unnecessary.

A typical example is the Euclidean metric:

$$d_2(x, y) = \sqrt{\sum_{j=1}^{p} (x^j - y^j)^2} = \|x - y\|_2$$

where $x = (x^1, \dots, x^p)$ and $y = (y^1, \dots, y^p)$ are p-dimensional vectors of $\boldsymbol{R}^p$. We sometimes use the Euclidean norm denoted by $\|x\|_2$ and the Euclidean scalar product $\langle x, y \rangle = \sum_{i=1}^{p} x^i y^i$. We note that $\|x\|_2^2 = \langle x, x \rangle$. Moreover we omit the subscript 2 and write $\|x\|$ instead of $\|x\|_2$ when no ambiguity arises.

Although the most frequently used space is the Euclidean space, what we use as a dissimilarity is not the Euclidean metric d_2 itself, but the squared metric:

$$D(x, y) = d_2^2(x, y) = \|x - y\|_2^2 = \sum_{j=1}^{p} (x^j - y^j)^2. \tag{2.3}$$

Note moreover that the word *distance* has been used in literature with the meaning of metric, whereas this word is informally used in this book.

A *cluster center* or *cluster prototype* is used in many algorithms for clustering; it is an element in the space $\boldsymbol{R}^p$ calculated as a function of elements in X.

The last remark in this section of terminology is that an *algorithm* for clustering implies not only a computational procedure but also a method of computation in a wider sense. In other words, the word of 'algorithm' is used rather informally and interchangeably with the word of 'method'.

Note 2.1.1. About the origin and rigorous definition of an algorithm, see, e.g., Knuth [88].

2.2 A Basic Algorithm of c-Means

A c-means algorithm classifies objects in X into c disjoint subsets G_i $(i = 1, 2, \dots, c)$ which are also called clusters. In each cluster, a cluster center v_i $(i = 1, 2, \dots, c)$ is determined.

We first describe a simple procedure which is most frequently used and quoted. Note that the phrases in the brackets are abbreviations.

Algorithm CM: *A Basic Procedure of Crisp c-Means.*

CM1. [Generate initial value:] Generate c initial values for cluster centers v_i $(i = 1, 2, \dots, c)$.

CM2. [Allocate to the nearest center:] Allocate all objects x_k $(k = 1, 2, \dots, N)$ to the cluster of the nearest center:

$$x_k \in G_i \iff i = \arg \min_{1 \le j \le c} D(x_k, v_j). \tag{2.4}$$

CM3. [Update centers:] Calculate new cluster centers as the centroids (center of gravity) of the corresponding clusters:

$$v_i = \frac{1}{|G_i|} \sum_{x_k \in G_i} x_k, \tag{2.5}$$

where $|G_i|$ is the number of elements in G_i, $i = 1, 2, \dots, c$.

CM4. [Test convergence:] If the clusters are convergent, stop; else go to CM2.

End CM.

Let us remark that the step CM4 has an informal description that *clusters are convergent.* We have two ways to judge whether or not clusters are convergent. That is, clusters are judged to be convergent when

(I) no object changes its membership from the last membership, or
(II) no centroid changes its position from the last position.

Note 2.2.1. We frequently use the term 'arg min' instead of 'min'. For example, $i = \arg \min_{1 \le j \le c} D(x_k, v_j)$ is the same as $D(x_k, v_i) = \min_{1 \le j \le c} D(x_k, v_j)$. Moreover we sometimes write like $v_i = \arg \min_{1 \le j \le c} D(x_k, v_j)$ by abuse of terminology when the last expression seems simplest.

2.3 Optimization Formulation of Crisp c-Means Clustering

Let $\boldsymbol{G} = (G_1, \dots, G_c)$ and $V = (v_1, \dots, v_c)$. The clusters $G_1, \dots, G_c$ are disjoint and their union is the set of objects:

$$\bigcup_{i=1}^{c} G_i = X, \qquad G_i \cap G_j = \emptyset \quad (i \neq j). \tag{2.6}$$

A sequence $\boldsymbol{G} = (G_1, \dots, G_c)$ satisfying (2.6) is called a *partition* of X.

Consider the following function having variables $\boldsymbol{G}$ and V.

$$J_{\rm cm}(\boldsymbol{G}, V) = \sum_{i=1}^{c} \sum_{x \in G_i} D(x, v_i), \tag{2.7}$$

where $D(x, v_i) = \|x - v_i\|_2^2$.

The function $J_{\rm cm}(\boldsymbol{G}, V)$ is the first objective function to be minimized for data clustering from which many considerations start. We will notice that to minimize an objective function for clustering is not an exact minimization but an iteration procedure of 'alternate minimization' with respect to a number of different variables.

Let us note correspondence between minimization of $J_{\rm cm}(\boldsymbol{G}, V)$ and a step in the procedure **CM**. We observe the following:

- the step CM2 (nearest center allocation) is equivalent to minimization of $J_{\rm cm}(\boldsymbol{G}, V)$ with respect to $\boldsymbol{G}$ while V is fixed;
- the step CM3 (to update centers) is equivalent to minimization of $J_{\rm cm}(\boldsymbol{G}, V)$ with respect to V while $\boldsymbol{G}$ is fixed.

To see the equivalence between the nearest allocation and $\min_{\boldsymbol{G}} J_{\rm cm}(\boldsymbol{G}, V)$, note that (2.4) minimizes $D(x_k, v_i)$ with respect to $i = 1, \dots, c$. Summing up $D(x_k, v_i)$ for all $1 \leq k \leq N$, we observe that the nearest allocation rule (2.4) realizes $\min_{\boldsymbol{G}} J_{\rm cm}(\boldsymbol{G}, V)$. The converse is also true by observing each x_k can be allocated to one of v_i's independently of other objects.

Next, its is well-known that the centroid v_i using (2.5) realizes

$$v_i = \arg\min_{v} \sum_{x \in G_i} \|x - v\|^2,$$

whence the equivalence between the calculation of centroid in CM3 and $\min_{V} J_{\rm cm}$ $(\boldsymbol{G}, V)$ is obvious.

Note 2.3.1. $\arg\min_{v}$ is an abbreviation of $\arg\min_{v \in \boldsymbol{R}^p}$. In this way, when no constraint is imposed on a variable, we omit the symbol of the universal set.

Observation of these equivalences leads us to the following procedure of iterative alternate optimization, which is equivalent to the first algorithm **CM**. Notice that the optimal solutions are denoted by $\bar{\boldsymbol{G}}$ and $\bar{V}$.

Algorithm CMO: *Crisp c-Means in Alternate Optimization Form.*

CMO1. [Generate initial value:] Generate c initial values for cluster centers $\bar{V} = (\bar{v}_1, \ldots, \bar{v}_c)$.

CMO2. Calculate

$$\bar{\boldsymbol{G}} = \arg\min_{\boldsymbol{G}} J_{\text{cm}}(\boldsymbol{G}, \bar{V}). \tag{2.8}$$

CMO3. Calculate

$$\bar{V} = \arg\min_{V} J(\bar{\boldsymbol{G}}, V). \tag{2.9}$$

CMO4. [Test convergence:] If $\bar{\boldsymbol{G}}$ or $\bar{V}$ is convergent, stop; else go to CMO2.

End CMO.

The convergence test using $\bar{\boldsymbol{G}}$ means that the new $\bar{\boldsymbol{G}}$ coincides with the last $\bar{\boldsymbol{G}}$; it is the same as no membership change in CM4. when $\bar{V}$ is used for the convergence test, it implies that the centroids do not change their positions.

Although **CMO** is not a new algorithm as it is equivalent to **CM**, we have the first mathematical result of convergence from this procedure (cf. [1]).

Proposition 2.3.1. *The algorithm* **CMO** *(and hence* **CM** *also) finally stops after a finite number of iterations of the major loop CMO2–CMO4 (resp. CM2–CM4).*

Proof. Let us assume that the convergence test uses $\bar{\boldsymbol{G}}$. As the procedure is iterative minimization, the value of objective function is monotonically non-increasing. As the number of all possible partitions of X is finite and the objective function is non-increasing, eventually the value of the objective function remains the same, which means that the membership of $\bar{\boldsymbol{G}}$ does not change.

The same argument is moreover valid when the convergence test uses $\bar{V}$. □

This property of convergence is actually very weak, since the number of all combinations for the partition $\boldsymbol{G}$ is huge. However, the observation of monotone non-increasing property stated in the proof will be used again in a number of variations of c-means algorithms described later. Moreover the monotone property suggests the third criterion for convergence: *stop the algorithm when the change of the value of the objective function is negligible.*

Let us proceed to consider a sequential algorithm [29], a variation of the basic c-means procedure. For this purpose note that updating cluster centers are done after all objects are reallocated to the nearest centers. An idea for the variation is to update cluster centers immediately after an object is moved from a cluster to another. A naïve algorithm using this idea is the following.

Algorithm CMS: *A Sequential Algorithm of Crisp c-Means*

CMS1. Generate c initial values for cluster centers v_i $(i = 1, 2, \ldots, c)$. Repeat CMS2 until convergence.

CMS2. Repeat CMS2-1 and CMS2-2 for all $x_k \in X$.

CMS2-1. [reallocate an object to the nearest center:] Assume $x_k \in G_j$. Calculate the nearest center to x_k:

$$i = \arg\min_{1 \le \ell \le c} D(x_k, v_\ell).$$

CMS2-2. [update centers:] If $i = j$, skip this step and go to CMS2-1. If $i \neq j$, update the centers:

$$\bar{v}_i = \frac{|G_i|}{|G_i|+1} v_i + \frac{x_k}{|G_i|+1} \tag{2.10}$$

$$\bar{v}_j = \frac{|G_j|}{|G_j|-1} v_j - \frac{x_k}{|G_j|-1} \tag{2.11}$$

in which $\bar{v}_i$ and $\bar{v}_j$ denote new centers, whereas v_i and v_j are old centers. Move x_k from G_j to G_i.

End CMS.

This simple procedure is based on the optimization of the same objective function $J_{\rm cm}(\boldsymbol{G}, V)$ but an object is moved at a time. Hence Proposition 2.3.1 with adequate modifications holds for algorithm **CMS**.

Another sequential algorithm has been proposed in [29]. Their algorithm uses objective function value as the criterion to change memberships:

Instead of CMS2-1, the next CMS'2-1 is used.

CMS'2-1. Assume x_k is in G_q. if $|G_q| \neq 1$,

$$j_\ell = \frac{|G_\ell|}{|G_\ell|+1} \|x_k - v_\ell\|^2, \qquad \ell \neq q$$

$$j_q = \frac{|G_q|}{|G_q|-1} \|x_k - v_q\|^2, \qquad \ell = 1, 2, \ldots, c.$$

If $j_r \leq j_\ell$ for all ℓ, and $r \neq q$, then move x_k from G_q to G_r. Update centroids:

$$v_q = v_q - \frac{x_k - v_q}{|G_q|-1}$$

$$v_r = v_r + \frac{x_k - v_r}{|G_r|+1}$$

and

$$|G_q| = |G_q| - 1, \quad |G_r| = |G_r| + 1$$

Update the objective function value:

$$J = J + j_r - j_q.$$

Use also the next CMS'3 instead of CMS3:

CMS'3. If the value of J in CMS'2-1 does not decrease, then stop. Else go to CMS'2-1.

In the last algorithm G_q and G_r change respectively to $G' = G_q - \{x_k\}$ and $G'' = G_r \cup \{x_k\}$.

Note 2.3.2. To see the above calculation is correct, note the follwoing holds

$$\sum_{x \in G'} \|x - M(G')\|^2 = \sum_{x \in G_q} \|x - M(G_q)\|^2 - \frac{|G_q|}{|G_q| - 1} \|x_k - M(G_q)\|^2$$

$$\sum_{x \in G''} \|x - M(G'')\|^2 = \sum_{x \in G_r} \|x - M(G_r)\|^2 + \frac{|G_r|}{|G_r| + 1} \|x_k - M(G_r)\|^2$$

where $M(G_q)$ and $M(G_r)$ are centroids of G_q and G_r, respectively.

Next consideration leads us to fuzzy c-means algorithms. For the most part discussion below to obtain fuzzy c-means algorithm is based on Bezdek [6]. For this purpose we write **CM** again using an additional variable of memberships.

Let us introduce a $N \times c$ membership matrix $U = (u_{ki})$ $(1 \le k \le N, 1 \le i \le c)$ in which u_{ki} takes a real value. Let us first assume u_{ki} is binary-valued ($u_{ki} = 0$ or 1) with the meaning that

$$u_{ki} = \begin{cases} 1 & (x_k \in G_i), \\ 0 & (x_k \notin G_i). \end{cases} \tag{2.12}$$

Thus each component of U shows membership/non-membership of an object to a cluster.

Then the objective function $J_{\text{cm}}(\boldsymbol{G}, V)$ can be rewritten using U:

$$J_0(U, V) = \sum_{i=1}^{c} \sum_{k=1}^{N} u_{ki} D(x_k, v_i) \tag{2.13}$$

This form alone is not equivalent to $J(\boldsymbol{G}, V)$, since $\boldsymbol{G}$ is a partition of X while u_{ki} does not exclude multiple belongingness of an object to more than one cluster.

We hence impose a constraint to U so that an object belongs to one and only one cluster: for all k, there exists a unique i such that $u_{ki} = 1$. This constraint can be expressed in the next form:

$$\begin{aligned} \mathcal{U}_b = \{ U = (u_{ki}) \, : \, & \sum_{j=1}^{c} u_{kj} = 1, \; 1 \le k \le N; \\ & u_{ki} \in \{0, 1\}, \; 1 \le k \le N, \; 1 \le i \le c \}. \end{aligned} \tag{2.14}$$

Using $\mathcal{U}_b$ steps CMO2 and CMO3 can be written as follows.

CMO2:

$$\bar{U} = \arg \min_{U \in \mathcal{U}_b} J_0(U, \bar{V}). \tag{2.15}$$

CMO3:

$$\bar{V} = \arg \min_{V} J_0(\bar{U}, V). \tag{2.16}$$

Notice that the solution of (2.16) is

$$\bar{v}_i = \frac{\sum_{k=1}^{c} \bar{u}_{ki} x_k}{\sum_{k=1}^{c} \bar{u}_{ki}}. \tag{2.17}$$

2.4 Fuzzy c-Means

We proceed to 'fuzzify' the constraint $\mathcal{U}_b$. That is, we allow fuzzy memberships by relaxing the condition $u_{ki} \in \{0, 1\}$ into the fuzzy one: $u_{ki} \in [0, 1]$. This means that we replace $\mathcal{U}_b$ by the next constraint:

$$\mathcal{U}_f = \{\, U = (u_{ki}) \, : \sum_{j=1}^{c} u_{kj} = 1, \, 1 \le k \le N; \\ u_{ki} \in [0, 1], \, 1 \le k \le N, \, 1 \le i \le c \,\}. \tag{2.18}$$

Apparently, the form of solution (2.17) for (2.16) remains the same even when fuzzy solution of U is used. In contrast, the solution

$$U = \arg \min_{U \in \mathcal{U}_f} J_0(U, \bar{V}). \tag{2.19}$$

appears to change in accordance with the generalization from $\mathcal{U}_b$ to $\mathcal{U}_f$. Nevertheless, we observe that the membership does not change; the solution $\bar{U}$ remains crisp.

Proposition 2.4.1. *The solution $\bar{U}$ of (2.19) is the same as that of (2.15).*

Proof. Notice that the objective function $J_0(U, V)$ is linear with respect to u_{ki}; the constraint (2.18) is also linear. Hence the fundamental property of linear programming which states that the optimal solution of a linear programming is at a vertex of the hyper-polygon of the feasible solutions is applicable. In the present case we have $\bar{u}_{ki} = 1$ or 0. Thus the solution can be assumed to be crisp, and hence the argument of the crisp minimization is used and we have the desired conclusion. □

This property implies that the use of $J_0(U, V)$ in an alternate minimization algorithm is useless when we wish to have a fuzzy solution. In other words, it is necessary to introduce nonlinearity for U to obtain fuzzy memberships.

For this purpose Bezdek [6] and Dunn [31, 32] introduced the nonlinear term $(u_{ki})^m$ into the objective function:

$$J_{\text{fcm}}(U, V) = \sum_{i=1}^{c} \sum_{k=1}^{N} (u_{ki})^m D(x_k, v_i), \quad (m > 1). \tag{2.20}$$

We will later introduce another type of nonlinearity to define a different method of fuzzy c-means clustering. To distinguish these different types, this algorithm by Dunn and Bezdek is called the standard method of fuzzy c-means. Let us describe again the basic alternate optimization procedure of fuzzy c-means for convenience.

Algorithm FCM: *Basic Procedure of Fuzzy c-Means.*

FCM1. [Generate initial value:] Generate c initial values for cluster centers $\bar{V} = (\bar{v}_1, \dots, \bar{v}_c)$.

FCM2. [Find optimal U:] Calculate

$$\bar{U} = \arg \min_{U \in \mathcal{U}_f} J_{\mathrm{fcm}}(U, \bar{V}). \tag{2.21}$$

FCM3. [Find optimal V:] Calculate

$$\bar{V} = \arg \min_{V} J_{\mathrm{fcm}}(\bar{U}, V). \tag{2.22}$$

FCM4. [Test convergence:] If $\bar{U}$ or $\bar{V}$ is convergent, stop; else go to FCM2.
End FCM.

Note 2.4.1. As a convergence criterion in FCM4, one of the followings can be used.

(i) For a small positive number ϵ, judge that the solution U is convergent if

$$\max_{k,i} |\bar{u}_{ki} - \hat{u}_{ki}| < \epsilon$$

where $\bar{U}$ is the new solution and $\hat{U}$ is the optimal solution one step before the last.

(ii) For a small positive number ϵ, judge that the solution V is convergent

$$\max_{1 \le i \le c} \|\bar{v}_i - \hat{v}_i\| < \epsilon$$

where $\bar{V}$ is the new solution and $\hat{V}$ is the optimal solution one step before the last.

(iii) Judge that the solution is convergent when the value of the objective function is convergent.

Besides one of these criteria, limitation on the maximum number of iterations of the major loop FCM2–FCM4 should be specified.

It is well-known that the solutions of FCM2 and FCM3 are respectively given by the following.

$$\bar{u}_{ki} = \left[\sum_{j=1}^{c} \left(\frac{D(x_k, \bar{v}_i)}{D(x_k, \bar{v}_j)} \right)^{\frac{1}{m-1}} \right]^{-1}, \tag{2.23}$$

$$\bar{v}_i = \frac{\sum_{k=1}^{N} (\bar{u}_{ki})^m x_k}{\sum_{k=1}^{N} (\bar{u}_{ki})^m}. \tag{2.24}$$

Note 2.4.2. The formula (2.23) exclude the case when $x_k = v_i$ for which we have the obvious solution $u_{ki} = 1$. Hence the exact formula is

$$\bar{u}_{ki} = \begin{cases} \left[\sum_{j=1}^{c} \left(\frac{D(x_k, \bar{v}_i)}{D(x_k, \bar{v}_j)} \right)^{\frac{1}{m-1}} \right]^{-1}, & (x_k \neq v_i), \\ 1 & (x_k = v_i). \end{cases} \tag{2.25}$$

We mostly omit the case of $\bar{u}_{ki} = 1$ $(x_k = v_i)$ and write the solutions simply by (2.23), as the omission is not harmful to most discussions herein.

Let us derive the solution (2.23). Since the constraint $\mathcal{U}_f$ has inequalities, it appears that the Kuhn-Tucker condition should be used. However, we have a simpler method to employ the conventional Lagrange multiplier for the equalities alone. We relax temporarily the constraint to

$$\left\{ (u_{ki}) : \sum_{i=1}^{c} u_{ki} = 1, \text{ all } k = 1, 2, \ldots, N \right\}$$

by removing

$$u_{ki} \geq 0, \quad 1 \leq k \leq N, 1 \leq i \leq c.$$

Suppose the solution $\tilde{U}$ by this relaxed condition satisfies the original constraint (i.e., $\tilde{U} \in \mathcal{U}_f$), then it is obvious that this solution is the optimal solution of the original problem ($\bar{U} = \tilde{U}$). To this end we consider minimization of J_{fcm} with respect to U under the condition $\sum_{i=1}^{c} u_{ki} = 1$ using the Lagrange multiplier.

Let the Lagrange multiplier be λ_k, $k = 1, \ldots, N$, and put

$$L = J_{\text{fcm}} + \sum_{k=1}^{N} \lambda_k \left(\sum_{i=1}^{c} u_{ki} - 1 \right)$$

For the necessary condition of optimality we differentiate

$$\frac{\partial L}{\partial u_{ki}} = m(u_{ki})^{m-1} D(x_k, v_i) + \lambda_k = 0.$$

We assume no v_i satisfies $x_k = v_i$. Then $D(x_k, v_j) > 0$ $(j = 1, \dots, c)$. To eliminate λ_k, we note

$$u_{kj} = \left[\frac{-\lambda_k}{mD(x_k, v_j)}\right]^{\frac{1}{m-1}} \tag{2.26}$$

Summing up for $j = 1, \dots, c$ and taking $\sum_{j=1}^{c} u_{kj} = 1$ into account, we have

$$\sum_{j=1}^{c} \left[\frac{-\lambda_k}{mD(x_k, v_j)}\right]^{\frac{1}{m-1}} = 1$$

Using (2.26) to this equation, we can eliminate λ_k, having

$$u_{ki} = \left[\sum_{j=1}^{c} \left(\frac{D(x_k, v_i)}{D(x_k, v_j)}\right)^{\frac{1}{m-1}}\right]^{-1}.$$

Since this solution satisfies $u_{ki} \geq 0$, it is the solution of FCM2. Moreover this solution is actually the unique solution of FCM2, since the objective function is strictly convex with respect to u_{ki}.

If there is v_i satisfying $x_k = v_i$, the optimal solution is obviously $u_{ki} = 1$ and $u_{kj} = 0$ for $j \neq i$.

The optimal solution (2.24) is easily derived by differentiating J_{fcm} with respect to V, since there is no constraint on V. We omit the detail.

Note 2.4.3. For the sake of simplicity, we frequently omit bars from the solutions $\bar{U}$ and $\bar{V}$ and write

$$u_{ki} = \left[\sum_{j=1}^{c} \left(\frac{D(x_k, v_i)}{D(x_k, v_j)}\right)^{\frac{1}{m-1}}\right]^{-1}, \tag{2.27}$$

$$v_i = \frac{\sum_{k=1}^{N} (u_{ki})^m x_k}{\sum_{k=1}^{N} (u_{ki})^m}, \tag{2.28}$$

when no ambiguity arises.

Apparently, the objective function converges to the crisp objective function as $m \to 1$:

$$J_{\text{fcm}}(U, V) \to J_0(U, V), \quad (m \to 1).$$

A question arises how the fuzzy solution $\bar{U}$ given by (2.23) is related to the crisp one. We have the next proposition.

Proposition 2.4.2. *As $m \to 1$, the solution (2.23) converges to the crisp solution (2.4), on the condition that the nearest center to any x_k is unique. In other words, for all x_k, there exists unique v_i such that*

$$i = \arg\min_{1 \le \ell \le c} D(x_k, v_\ell).$$

Proof. Note

$$\frac{1}{u_{ki}} - 1 = \sum_{j \ne i} \left(\frac{D(x_k, v_i)}{D(x_k, v_j)} \right)^{\frac{1}{m-1}}.$$

Assume v_i is nearest to x_k. Then all terms in the right hand side are less than unity. Hence the right hand side tends to zero as $m \to 1$. Assume v_i is not nearest to x_k. Then a term in the right hand side exceeds unity. Hence the right hand side tends to $+\infty$ as $m \to 1$. The proposition is thus proved. □

For a discussion later, we introduce a function

$$F_{DB}(x; v_j) = \frac{1}{D(x, v_j)^{\frac{1}{m-1}}}. \tag{2.29}$$

We then have

$$u_{ki} = \frac{F_{DB}(x_k; v_i)}{\displaystyle\sum_{j=1}^{c} F_{DB}(x_k; v_j)}. \tag{2.30}$$

Note 2.4.4. The constraint $\mathcal{U}_f$ is a compact set of $\boldsymbol{R}^p$ and hence optimal solution of a continuous function takes its minimum value in $\mathcal{U}_f$. This fact is not very useful in the above discussion, since the function is convex and hence it has a unique global minimum. However, it is generally necessary to observe a topological property of the set of feasible solutions when we study a more complex optimization problem.

2.5 Entropy-Based Fuzzy c-Means

We first generalize algorithm **FCM** to **FC**$(J, \mathcal{U})$ which has two arguments of an objective function J and a constraint $\mathcal{U}$:

Algorithm FC$(J, \mathcal{U})$**:** *Generalized procedure of fuzzy c-means with arguments.*

FC1. [Generate initial value:] Generate c initial values for cluster centers $\bar{V} = (\bar{v}_1, \ldots, \bar{v}_c)$.

FC2. [Find optimal U:] Calculate

$$\bar{U} = \arg\min_{U \in \mathcal{U}} J(U, \bar{V}). \tag{2.31}$$

FC3. [Find optimal V:] Calculate

$$\bar{V} = \arg\min_{V} J(\bar{U}, V). \tag{2.32}$$

FC4. [Test convergence:] If $\bar{U}$ or $\bar{V}$ is convergent, stop; else go to FC2.
End FC.

Thus $\mathbf{FC}(J_0, \mathcal{U}_b)$ is employed for crisp c-means; $\mathbf{FC}(J_{\mathrm{fcm}}, \mathcal{U}_f)$ for fuzzy c-means.

We consider if there are other ways to fuzzify the crisp c-means. As the standard method by Dunn and Bezdek introduces nonlinearity $(u_{ki})^m$, we should consider the use of another type of nonlinearity.

The method of Dunn and Bezdek has another feature: it smoothes the crisp solution into a differentiable one. Moreover the fuzzy solution approximates the crisp one in the sense that the fuzzy solution converges to the crisp solution as $m \to 1$. Roughly, we can say that the fuzzified solution 'regularizes' the crisp solution.

Such an idea of regularization has frequently been found in the formulation of ill-posed problems (e.g.,[154]). A typical regularization is done by adding a regularizing function. In the present context, we consider

$$J'(U,V) = J_0(U,V) + \nu K(u), \quad (\nu > 0) \tag{2.33}$$

in which $K(u)$ is a nonlinear regularizing function and ν is a regularizing parameter.

We study two types of the regularizing function: one is an entropy function and another is quadratic.

$$K(u) = \sum_{i=1}^{c} \sum_{i=1}^{N} u_{ki} \log u_{ki}, \tag{2.34}$$

$$K(u) = \frac{1}{2} \sum_{i=1}^{c} \sum_{i=1}^{N} u_{ki}^2 \tag{2.35}$$

Notice that both functions are strictly convex function, and hence are capable of fuzzifying the membership matrix. When J' is used in algorithm **FC**, the both method guarantees unique solutions in FC2.

When we use the former, the algorithm is called regularization by entropy or an entropy-based method. The first formulation of this idea is maximum entropy method by Li and Mukaidono [90]; later Miyamoto and Mukaidono [102] have reformulated using the idea of regularization.

Namely, the following objective function is used for the entropy-based method.

$$J_{\mathrm{efc}}(U,V) = \sum_{i=1}^{c} \sum_{k=1}^{N} u_{ki} D(x_k, v_i) + \nu \sum_{i=1}^{c} \sum_{k=1}^{N} u_{ki} \log u_{ki}, \quad (\nu > 0). \tag{2.36}$$

Thus the method of entropy uses the algorithm $\mathbf{FC}(J_{\mathrm{efc}}, \mathcal{U}_f)$. The solutions in the steps FC2 and FC3 are as follows.

$$u_{ki} = \frac{\exp\left(-\frac{D(x_k, v_i)}{\nu}\right)}{\sum_{j=1}^{c} \exp\left(-\frac{D(x_k, v_j)}{\nu}\right)}, \tag{2.37}$$

$$v_i = \frac{\sum_{k=1}^{N} u_{ki} x_k}{\sum_{k=1}^{N} u_{ki}}. \tag{2.38}$$

In parallel to Proposition 2.4.2, we have

Proposition 2.5.1. *As $\nu \to 0$, the solution (2.37) converges to the crisp solution (2.4), on the condition that the nearest center to any x_k is unique. That is, for all x_k, there exists unique v_i such that $i = \arg\min_{1 \le \ell \le c} D(x_k, v_\ell)$.*

Proof. Note

$$\frac{1}{u_{ki}} - 1 = \sum_{j \neq i} \exp\left(\frac{D(x_k, v_j) - D(x_k, v_i)}{\nu}\right).$$

Assume v_i is nearest to x_k. Then all terms in the right hand side are less than unity. Hence the right hand side tends to zero as $\nu \to 0$. Assume v_i is not nearest to x_k. Then a term in the right hand side exceeds unity. Hence the right hand side tends to $+\infty$ as $\nu \to 0$. The proposition is thus proved. □

In parallel to $F_{DB}(x; v_i)$ by (2.29), we introduce a function

$$F_E(x; v_i) = \exp\left(-\frac{D(x, v_i)}{\nu}\right). \tag{2.39}$$

We then have

$$u_{ki} = \frac{F_E(x_k; v_i)}{\sum_{j=1}^{c} F_E(x_k; v_j)}. \tag{2.40}$$

Note 2.5.1. Let us relax the constraint $\sum_i u_{ki} = 1$, i.e., $0 \le u_{ki} \le 1$ is deleted as before. Put $D_{ki} = D(x_k, v_i)$ for simplicity. Define the Lagrangian

$$L = \sum_{k=1}^{N} \sum_{i=1}^{c} u_{ki} D_{ki} + \nu \sum_{k=1}^{N} \sum_{i=1}^{c} u_{ki} \log u_{ki} + \sum_{k=1}^{N} \lambda_k (\sum_{i=1}^{c} u_{ki} - 1).$$

where λ_k is the Lagrange multipliers. From

$$\frac{\partial L}{\partial u_{ki}} = D_{ki} + \nu(1 + \log u_{ki}) + \lambda_k = 0,$$

we have

$$u_{ki} = \exp(-D_{ki}/\nu - \lambda_k/\nu - 1). \tag{2.41}$$

Note that the above solution satisfies $0 \le u_{ki} \le 1$. Hence (2.41) provides the solution for the stationary point under the constraint $\mathcal{U}_f$. Since λ_k from $\sum_i u_{ki} = 1$ we have

$$\sum_i u_{kj} = \exp(-\lambda_k/\nu - 1)\sum_j \exp(-D_{kj}/\nu) = 1.$$

Eliminating $\exp(-\lambda_k/\nu - 1)$ from the both equations, we have (2.37). Since the objective function is strictly convex with respect to u_{ki}, we note the optimal solution is (2.37).

Note 2.5.2. Maximum entropy principle has been studied in different fields (e.g., [170]). The method of Li and Mukaidono was maximum entropy method with the equality constraint:

$$\text{maximize} \quad -\sum_{i=1}^{c}\sum_{k=1}^{N} u_{ki}\log u_{ki}$$

$$\text{subject to} \quad \sum_{i=1}^{c}\sum_{k=1}^{N} u_{ki}D_{ki} = L.$$

It is easily seen that this formulation is equivalent to the regularization by entropy when the value L is not given but may be changed as a parameter. Namely, unknown L may be replaced by the regularizing parameter ν, while ν is interpreted as the Lagrange multiplier in the maximum entropy method (cf. [154]). Our idea is to use the concept of regularization which we believe is more adequate to fuzzy clustering. See also [96] for similar discussion.

2.6 Addition of a Quadratic Term

We consider the second method of adding the quadratic term. That is, the next objective function is considered:

$$J_{\rm qfc}(U,V) = \sum_{i=1}^{c}\sum_{k=1}^{N} u_{ki}D(x_k,v_i) + \frac{1}{2}\nu\sum_{i=1}^{c}\sum_{k=1}^{N} u_{ki}^2 \quad (\nu > 0), \tag{2.42}$$

and algorithm $\mathbf{FC}(J_{\rm qfc},\mathcal{U}_f)$ is used.

The solution for FC3 is given by (2.38) which is the same as that for the entropy method. In contrast, the solution in the step FC2 does not have a simple form such as (2.23) and (2.37); it is given by an algorithm which needs a complicated derivation process. Readers uninterested in the quadratic method may skip this section.

2.6.1 Derivation of Algorithm in the Method of the Quadratic Term

We derive the necessary condition for the optimal solution. For this purpose we first put $w_{ki}^2 = u_{ki}$. Then $u_{ki} \geq 0$ is automatically satisfied and the constraint is represented by $\sum_i w_{ki}^2 = 1$.

We hence define the Lagrangian

$$L = \sum_{k=1}^{N} \sum_{i=1}^{c} w_{ki}^2 D_{ki} + \frac{1}{2}\nu \sum_{k=1}^{N} \sum_{i=1}^{c} w_{ki}^4 - \sum_{k=1}^{N} \mu_k (\sum_{i=1}^{c} w_{ki}^2 - 1)$$

where $D_{ki} = D(x_k, v_i)$.

From

$$\frac{1}{2} \frac{\partial L}{\partial w_{ki}} = w_{ki}(D_{ki} + \nu w_{ki}^2 - \mu_k) = 0,$$

we have $w_{ki} = 0$ or $w_{ki}^2 = \nu^{-1}(\mu_k - D_{ki})$. Using u_{ki},

$$u_{ki} = 0 \quad \text{or} \quad u_{ki} = \nu^{-1}(\mu_k - D_{ki}). \tag{2.43}$$

Notice that $u_{ki} = \nu^{-1}(\mu_k - D_{ki}) \geq 0$.

The above solution has been derived from the necessary condition for optimality. We hence should find the optimal solution from the set of u_{ki} that satisfies (2.43).

Let us simplify the problem in order to find the optimal solution. Let

$$J^{(k)}(U, v) = \sum_{i=1}^{c} u_{ki} D_{ki} + \frac{1}{2}\nu \sum_{i=1}^{c} u_{ki}^2 \tag{2.44}$$

Then, $J_{\text{qfc}} = \sum_{k=1}^{N} J^{(k)}$ and each $J^{(k)}$ can independently be minimized from other $J^{(k')}$ $(k' \neq k)$. Hence we consider optimization of $J^{(k)}$ for a fixed k. Assume

$$D_{k,k_1} \leq \cdots \leq D_{k,k_c} \tag{2.45}$$

and we rewrite

$$D_{k1} \leq D_{k2} \leq \cdots \leq D_{kc} \tag{2.46}$$

instead of (2.45) for simplicity by renumbering clusters.

Suppose that we have found u_{ki}, $i = 1, \ldots, c$, that satisfies the condition (2.43). Let I be a set of indices such that u_{ki} is positive, i.e.,

$$I = \{i : u_{ki} > 0\}, \tag{2.47}$$

and $|I|$ be the number of elements in I. From (2.43) and $\sum_i u_{ki} = 1$, we have

$$\nu^{-1}|I|\mu_k - \nu^{-1} \sum_{\ell \in I} D_{k\ell} = 1. \tag{2.48}$$

This implies that, for $i \in I$,

$$u_{ki} = |I|^{-1}(\nu^{-1} \sum_{\ell \in I} D_{k\ell} + 1) - \nu^{-1} D_{ki} > 0 \tag{2.49}$$

while $u_{ki} = 0$ for $i \notin I$.

The problem is thus reduced to the optimal choice of the index set I, since when I is found, it is easy to calculate the membership values by (2.49). It has been proved that the optimal choice of the index set has the form of $I = \{1, \dots, K\}$ for some $K \in \{1, 2, \dots, c\}$. Moreover let

$$f_\ell^L = \nu^{-1}(\sum_{j=1}^{L}(D_{kj} - D_{k\ell})) + 1, \quad \ell = 1, \dots, L. \tag{2.50}$$

Whether or not f_L^L is positive determines the optimal index set. The detail of the proof is found in [104], which is omitted here, and the final algorithm is given in the following.

Algorithm for the optimal solution of u_{ki}

Under the assumption (2.46), the solution that minimizes $J^{(k)}$ is given by the following procedure.

1. Calculate f_L^L for $L = 1, \dots, c$ by (2.50). Let $\bar{L}$ be the smallest number such that $f_{\bar{L}+1}^{\bar{L}+1} \le 0$. Namely, $I = \{1, 2, \dots, \bar{L}\}$ is the optimal index set.
2. For $i = 1, \dots, \bar{L}$, put

$$u_{ki} = \bar{L}^{-1}(\nu^{-1} \sum_{\ell=1}^{\bar{L}} D_{k\ell} + 1) - \nu^{-1} D_{ki}$$

and for $i = \bar{L} + 1, \dots, c$, put $u_{ki} = 0$.

2.7 Fuzzy Classification Rules

Suppose that we clustered given objects X using a crisp or fuzzy c-means algorithm, and after clustering a new element $x \in \boldsymbol{R}^p$ is observed. A question is: 'in what way should we determine the cluster to which x belongs?' Although we might say no unique solution exists, we can find a natural and adequate answer to this problem.

To this end, we note that each method of clustering has its intrinsic allocation rule. In the crisp c-means the rule is allocation to the nearest center in CM2. For classifying x, the nearest allocation rule is

$$x \in G_i \iff i = \arg \min_{1 \le j \le c} D(x, v_j). \tag{2.51}$$

When $D(x, v_i) = D(x, v_j)$ for the two centers, we cannot determine the class of x. We can move x arbitrarily in $\boldsymbol{R}^p$ and accordingly the whole space is divided into c classes $\mathcal{H}(V)$:

$$
\begin{aligned}
&\mathcal{H}(V) = \{H_1, \dots, H_c\} \\
&H_i = \{\, x \in \boldsymbol{R}^p \; : \; \|x - v_i\| < \|x - v_\ell\|, \; \forall \ell \neq i \,\}
\end{aligned}
\tag{2.52}
$$

Geometrically the c classes in $\boldsymbol{R}^p$ are called the Voronoi sets [85] with centers v_i $(i = 1, \dots, c)$.

Next question is what the classification rules in the methods of fuzzy c-means are. Naturally an allocation rule should be a fuzzy rule also called a *fuzzy classifier.*

Here we use a term of a *fuzzy classification function* for the present fuzzy rule, since different ideas have been proposed under the name of a fuzzy classifier and we wish to avoid confusion.

Let us note the two functions $F_{DB}(x; v_i)$ and $F_E(x; v_i)$. Put

$$
\mathbf{U}^{(i)}_{\text{fcm}}(x; V) = \begin{cases} \dfrac{F_{DB}(x; v_i)}{\sum_{j=1}^{c} F_{DB}(x; v_j)}, & (x \neq v_i) \\ 1 & (x = v_i) \end{cases}
\tag{2.53}
$$

$$
\mathbf{U}^{(i)}_{\text{efc}}(x; V) = \frac{F_E(x; v_i)}{\sum_{j=1}^{c} F_E(x; v_j)} .
\tag{2.54}
$$

These classification functions are actually used in the respective clustering algorithms, since the membership is obtained by putting $x = x_k$, i.e., $\bar{u}_{ki} = \mathbf{U}^{(i)}_{\text{fcm}}(x_k; V)$ in **FCM** and $\bar{u}_{ki} = \mathbf{U}^{(i)}_{\text{efc}}(x_k; V)$ in $\mathbf{FC}(J_{\text{efc}}, \mathcal{U}_f)$.

Thus the two functions $\mathbf{U}^{(i)}_{\text{fcm}}(x; V)$ and $\mathbf{U}^{(i)}_{\text{efc}}(x; V)$ are the fuzzy classification functions to represent the fuzzy allocation rules. These functions are defined on the whole space: $\mathbf{U}^{(i)}_{\text{fcm}}(\cdot; V) \colon \boldsymbol{R}^p \to [0, 1]$; $\mathbf{U}^{(i)}_{\text{efc}}(\cdot; V) \colon \boldsymbol{R}^p \to [0, 1]$. The classification function forms a partition of the whole space:

$$
\begin{aligned}
&\sum_{j=1}^{c} \mathbf{U}^{(j)}_{\text{fcm}}(x; V) = 1, \quad \forall x \in \boldsymbol{R}^p, \\
&\sum_{j=1}^{c} \mathbf{U}^{(j)}_{\text{efc}}(x; V) = 1, \quad \forall x \in \boldsymbol{R}^p.
\end{aligned}
$$

Frequently fuzzy clusters are made crisp by the reallocation of x_k to G_i by the maximum membership rule:

$$
x_k \in G_i \iff u_{ki} = \max_{1 \le j \le c} u_{kj}.
\tag{2.55}
$$

It is easy to see that this reallocation rule classifies $\boldsymbol{R}^p$ into Voronoi sets, since it uses the maximum membership rule:

$$
x \in G_i \iff \mathbf{U}^{(i)}_{\text{fcm}}(x; V) > \mathbf{U}^{(j)}_{\text{fcm}}(x; V), \; \forall j \neq i \quad (x \in \boldsymbol{R}^p)
$$

or equivalently,

$$x \in G_i \iff \|x - v_i\| < \|x - v_j\|, \ \forall j \neq i \quad (x \in \boldsymbol{R}^p).$$

The same property holds for $\mathbf{U}^{(i)}_{\text{efc}}(x; V)$. Thus the crisp reallocation rules are the same as the classification rule of the crisp c-means.

We next note a number of properties of the classification functions whereby we have insight to characteristics of different methods of fuzzy c-means.

Proposition 2.7.1. *$\mathbf{U}^{(i)}_{\text{fcm}}(x; V)$ and $\mathbf{U}^{(i)}_{\text{efc}}(x; V)$ are continuous on $\boldsymbol{R}^p$.*

Proof. It is sufficient to check that $\mathbf{U}^{(i)}_{\text{fcm}}(x; V)$ is continuous at $x = v_i$. From

$$\begin{aligned} 1/\mathbf{U}^{(i)}_{\text{fcm}}(x; V) - 1 &= \sum_{j \neq i} \frac{F_{DB}(x; v_j)}{F_{DB}(x; v_i)} \\ &= \sum_{j \neq i} \left(\frac{\|x - v_i\|}{\|x - v_j\|} \right)^{\frac{2}{m-1}} \end{aligned} \tag{2.56}$$

we have

$$1/\mathbf{U}^{(i)}_{\text{fcm}}(x; V) - 1 \to 0 \quad \text{as} \quad x \to v_i$$

whence $\lim_{x \to v_i} \mathbf{U}^{(i)}_{\text{fcm}}(x; V) = 1$. It is also obvious that the second function $\mathbf{U}^{(i)}_{\text{efc}}(x; V)$ is continuous on $\boldsymbol{R}^p$. □

Proposition 2.7.2. *The function $\mathbf{U}^{(i)}_{\text{fcm}}(x; V)$ takes its maximum value 1 at $x = v_i$ while it tends to $1/c$ as $\|x\| \to +\infty$:*

$$\max_{x \in \boldsymbol{R}^p} \mathbf{U}^{(i)}_{\text{fcm}}(x; V) = \mathbf{U}^{(i)}_{\text{fcm}}(v_i; V) = 1 \tag{2.57}$$

$$\lim_{\|x\| \to +\infty} \mathbf{U}^{(i)}_{\text{fcm}}(x; V) = \frac{1}{c}. \tag{2.58}$$

Proof. Using (2.56) we immediately observe that

$$1/\mathbf{U}^{(i)}_{\text{fcm}}(x; V) - 1 > 0$$

for all $x \neq v_i$, while $\mathbf{U}^{(i)}_{\text{fcm}}(v_i; V) = 1$, from which the first property follows.

For the second property we can use (2.56) again:

$$1/\mathbf{U}^{(i)}_{\text{fcm}}(x; V) - 1 \to c - 1 \quad (\|x\| \to +\infty).$$ □

In contrast to the classification function $\mathbf{U}^{(i)}_{\text{fcm}}(x; V)$ for the standard method of fuzzy c-means, $\mathbf{U}^{(i)}_{\text{efc}}(x; V)$ for the entropy-based method has different characteristics related to a feature of Voronoi sets.

A Voronoi set H is bounded when H is included into a sufficiently large cube of $\boldsymbol{R}^p$, or it is unbounded when such a cube does not exist. When the entropy-based method is used, whether a Voronoi set is bounded or unbounded is determined by the configuration of $v_1, \ldots, v_c$. It is not difficult to see the following property is valid.

Property 2.7.1. A Voronoi set H with the center v_i determined by the centers $V = (v_1, \dots, v_c)$ is unbounded if and only if there exists a hyperplane in $\boldsymbol{R}^p$ such that $v_j - v_i$, $\forall\, j \neq i$, is on one side of the hyperplane.

The proof of this property is not difficult and omitted here. We now have the next proposition.

Proposition 2.7.3. *Assume that a Voronoi set H with center v_i is unbounded and there is no hyperplane stated in Property 2.7.1 on which three or more centers are on that hyperplane. Then*

$$\lim_{\|x\| \to \infty} \mathbf{U}^{(i)}_{\text{efc}}(x; V) = 1, \tag{2.59}$$

where x moves inside H. On the other hand, if H is bounded, we have

$$\max_{x \subset H} \mathbf{U}^{(i)}_{\text{efc}}(x; V) < 1. \tag{2.60}$$

Proof. Note

$$\begin{aligned} 1/\mathbf{U}^{(i)}_{\text{efc}}(x; V) - 1 &= \sum_{j \neq i} \frac{F_E(x; v_j)}{F_E(x; v_i)} \\ &= \sum_{j \neq i} e^{\frac{\|x - v_j\|^2 - \|x - v_i\|^2}{\nu}} \\ &= Const \cdot \sum_{j \neq i} e^{\frac{2\langle x, v_j - v_i \rangle}{\nu}} \end{aligned}$$

where $Const$ is a positive constant. x can be moved to infinity inside H such that $\langle x, v_j - v_i \rangle$ is negative for all $j \neq i$. Hence the right hand side of the above equation approaches zero and we have the first conclusion. The second conclusion is easily seen from the above equation. We omit the detail. □

Notice that the condition excluded in Proposition 2.7.3 'there exists a hyperplane stated in Property 2.7.1 on which three or more centers are on that hyperplane' is exceptional.

Next proposition states that $\mathbf{U}^{(i)}_{\text{efc}}(x; V)$ is a convex fuzzy set.

Proposition 2.7.4. $\mathbf{U}^{(i)}_{\text{efc}}(x; V)$ *is a convex fuzzy set of $\boldsymbol{R}^p$. In other words, all α-cut*

$$[\mathbf{U}^{(i)}_{\text{efc}}(x; V)]_\alpha = \{\, x \in \boldsymbol{R}^p \,:\, \mathbf{U}^{(i)}_{\text{efc}}(x; V) \geq \alpha \,\}, \quad (0 \leq \alpha < 1)$$

is a convex set of $\boldsymbol{R}^p$.

Proof. It is easy to see that $e^{\frac{2\langle x, v_j - v_i \rangle}{\nu}}$ is a convex function and finite sum of convex functions is convex. Hence $1/\mathbf{U}^{(i)}_{\text{efc}}(x; V)$ is also a convex function which means that any level set of $1/\mathbf{U}^{(i)}_{\text{efc}}(x; V)$ is convex. Thus the proposition is proved. □

The classification function for the method of quadratic term can also be derived but we omit the detail (see [104, 110]).

2.8 Clustering by Competitive Learning

The algorithms stated so far are based on the idea of the alternate optimization. In contrast, the paradigm of learning [85, 30] is also popular and useful as a basic idea for clustering data. In this section we introduce a basic clustering algorithm based on competitive learning.

Competitive learning algorithms are also related to the first method of c-means. To see the relation, we show an abbreviated description of the procedure **CM**, which is shown as C1–C4.

C1. Generate initial cluster centers.
C2. Select an object (or a set of objects) and allocate to the cluster of the nearest center.
C3. Update the cluster centers.
C4. If the clusters are convergent, stop; else go to C2.

Notice that these descriptions are abstractions of CM1–CM4.

The use of cluster centers and the nearest center allocation rule are common to various clustering algorithms and difficult to change unless a certain model such as fuzzy or statistical model is assumed. In contrast, updating cluster centers is somewhat arbitrary. No doubt the centroid — the center of gravity — is a standard choice, but other choice is also possible.

The original c-means and other algorithms stated so far are based on the fundamental idea of optimization. Optimization is a fundamental idea found in physics and economics and a huge number of studies has been done on optimization and a deep theory has been developed. The paradigm of learning is more recent and now is popular. In many applications methods based on learning are as effective as those based on optimization, and sometimes the former outperforms the latter.

We show a typical and simple algorithm of clustering using learning referring to Kohonen's works [85]. Kohonen shows the LVQ (Learning Vector Quantization) algorithm. Although LVQ is a method of supervised classification and not a clustering algorithm [85], modification of LVQ to a simple clustering algorithm is straightforward.

In many methods based on learning, an object is randomly picked from X to be clustered. Such randomness is inherent to a learning algorithm which is fundamentally different from optimization algorithms.

Algorithm LVQC: *clustering by LVQ.*

LVQC1. Set an initial value for $m_i, i = 1, \dots, c$ (for example, select c objects randomly as m_i, $i = 1, \dots, c$).
LVQC2. For $t = 1, 2, \dots,$ repeat LVQC3–LVQC5 until convergence (or until the maximum number of iterations is attained).
LVQC3. Select randomly $x(t)$ from X.
LVQC4. Let

$$m_l(t) = \arg \min_{1 \le i \le c} \|x(t) - m_i(t)\|.$$

LVQC5. Update $m_1(t), \ldots, m_c(t)$:

$$m_l(t+1) = m_l(t) + \alpha(t)[x(t) - m_l(t)],$$
$$m_i(t+1) = m_i(t), \quad i \neq l.$$

Object represented by $x(t)$ is allocated to G_l.
End LVQC.

In this algorithm, the parameter $\alpha(t)$ satisfies

$$\sum_{t=1}^{\infty} \alpha(t) = \infty, \ \sum_{t=1}^{\infty} \alpha^2(t) < \infty, \quad t = 1, 2, \cdots$$

For example, $\alpha(t) = Const/t$ satisfies these conditions.

The *codebook vectors* are thus cluster centers in this algorithm. The nearest center allocation is done in LVQC2 while the center $m_i(t)$ is gradually *learning* its position in the step LVQC5.

Note 2.8.1. The above algorithm LVQ is sometimes called VQ (vector quantization) which is an approximation of a probability density by a finite number of codebook vectors [85]. We do not care much about the original name. When used for supervised classification, a class is represented by more than one codebook vectors, while only one vector is used as a center for a cluster in unsupervised classification.

2.9 Fixed Point Iterations – General Consideration

There are other classes of algorithms of fuzzy clustering that are related to fuzzy c-means. Unlike the above stated methods, a precise characterization of other algorithms are difficult, since they have heuristic or *ad hoc* features. However, there is a broad framework which encompasses most algorithms, which is called *fixed point iteration.*

We briefly (and rather roughly) note a general method of fixed point iteration. Let S be a compact subset of $\boldsymbol{R}^p$ and T be a mapping defined on S into S ($T\colon S \to S$). An element $x \in S$ is said to be a fixed point of T if and only if

$$T(x) = x.$$

There have been many studies on fixed points, and under general conditions the existence of a fixed point has been proved (see Note 2.9.1 below).

Suppose that we start from an initial value $x^{(1)}$ and iterate

$$x^{(n+1)} = T(x^{(n)}), \quad n = 1, 2, \ldots \tag{2.61}$$

When a fixed point $\tilde{x}$ exists and we expect the iterative solution converges to the fixed point, the iterative calculation is called fixed point iteration.

Let us use an abstract symbols and write (2.23) and (2.24) respectively by $\bar{U} = T_1(U, \bar{V})$, $\bar{V} = T_2(\bar{U}, V)$. When the number of iterations is represented by

$n = 1, 2, \ldots$ and the solution of the n-th iteration is expressed as $U^{(n)}$ and $V^{(n)}$, then the above form is rewritten as

$$U^{(n+1)} = T_1(U^{(n)}, V^{(n)}), \quad V^{(n+1)} = T_2(U^{(n+1)}, V^{(n)}). \tag{2.62}$$

Although not exactly the same as (2.61), this iterative calculation procedure is a variation of fixed point iteration. Notice that when the solution $(\bar{U}, \bar{V})$ is convergent, the result is the fixed point of (2.62).

Note that $u_{ki} \in [0, 1]$ and v_i is in the convex hull with the vertices of X, the solution is in a rectangle of $\boldsymbol{R}^{c \times N + c \times p}$. Hence the mapping in (2.62) has a fixed point from Brouwer's theorem (cf. Note 2.9.1).

However, the iterative formula (2.62), or (2.23) and (2.24) in the original form, does not necessarily converge, as conditions for convergence to fixed points are far stronger than those for the existence (see Note 2.9.2).

Thus an iterative formula such as (2.62) is not necessarily convergent. However, experiences in many simulations and real applications tell us that such algorithms as described above actually converge, and even in worst cases when we observe divergent solutions we can change parameters and initial values and try again.

Note 2.9.1. There have been many fixed point theorems. The most well-known is Brouwer's theorem which state that any continuous function $T\colon S \to S$ has a fixed point when S is homeomorphic to a closed ball of $\boldsymbol{R}^p$ (see e.g., [158]).

Note 2.9.2. In contrast to the existence of a fixed point, far stronger conditions are needed to guarantee the convergence of a fixed point iteration algorithm. A typical condition is a contraction mapping which means that

$$\|T(x) - T(y)\| \le Const\|x - y\|$$

for all $x, y \in S$, where $0 < Const < 1$. In this case the iteration

$$x^{(n+1)}) = T(x^{(n)})), \quad n = 1, 2, \ldots$$

leads to $x^{(n)} \to \tilde{x}$ and $\tilde{x}$ is the fixed point $T(\tilde{x}) = \tilde{x}$.

Note 2.9.3. Apart from fixed point iterations, the convergence of fuzzy c-means solutions has been studied [6], but no strong result of convergence has been proved up to now.

2.10 Heuristic Algorithms of Fixed Point Iterations

Many heuristic algorithms that are not based on optimization can be derived. As noted above, we call them fixed point iterations. Some basic methods are described here, while others will be discussed later when we show variations, in particular those including covariance matrix.

We observe there are at least two ideas for fixed point iterations. One is the combination of a non-Euclidean dissimilarity and centroid calculation. This idea is to use the basic fuzzy c-means algorithm even for non-Euclidean dissimilarity although the solution of v_i for a non-Euclidean dissimilarity does not minimize the objective function. Let $D'(x_k, v_i)$ be a dissimilarity not necessarily Euclidean. The following iterative calculation is used.

$$u_{ki} = \left[\sum_{j=1}^{c} \left(\frac{D'(x_k, v_i)}{D'(x_k, v_j)} \right)^{\frac{1}{m-1}} \right]^{-1}, \tag{2.63}$$

$$v_i = \frac{\sum_{k=1}^{N} (u_{ki})^m x_k}{\sum_{k=1}^{N} (u_{ki})^m}. \tag{2.64}$$

In the case of the Euclidean dissimilarity $D'(x_k, v_i) = \|x_k - v_i\|^2$, (2.63) and (2.64) are respectively the same as (2.23) and (2.24), and the alternate optimization is attained.

We can also use the entropy-based method:

$$u_{ki} = \frac{\exp\left(-\frac{D'(x_k, v_i)}{\nu}\right)}{\sum_{j=1}^{c} \exp\left(-\frac{D'(x_k, v_j)}{\nu}\right)}, \tag{2.65}$$

$$v_i = \frac{\sum_{k=1}^{N} u_{ki} x_k}{\sum_{k=1}^{N} u_{ki}}. \tag{2.66}$$

As a typical example of non-Euclidean dissimilarity, we can mention the Minkowski metric:

$$D'_q(x, v) = \left[\sum_{\ell=1}^{p} (x^\ell - v^\ell)^q \right]^{\frac{1}{q}}, \quad q \geq 1. \tag{2.67}$$

Notice that x^ℓ is the ℓ's component of vector x. When $q = 1$, the Minkowski metric is called the L_1 metric or city-block metric.

Notice that (2.63) and (2.65) respectively minimize the corresponding objective function, while (2.64) and (2.66) do not.

Later we will show an exact and simple optimization algorithm for v_i in the case of the L_1 metric, while such algorithm is difficult to obtain when $q \neq 1$.

Another class of fixed point algorithms uses combination of a Euclidean or non-Euclidean dissimilarity and learning for centers. Let $D(x_k, v_i)$ be such a dissimilarity. We can consider the following algorithm that is a mixture of **FCM** and **LVQ**.

Algorithm FLC: *Clustering by Fuzzy Learning.*

FLC1. Set initial value for m_i, $i = 1, \dots, c$ (for example, select c objects randomly as $m_i, i = 1, \dots, c$).
FLC2. For $t = 1, 2, \dots$ repeat FLC3–FLC5 until convergence or maximum number of iterations.
FLC3. Select randomly $x(t)$ from X.
FLC4. Let

$$u_{ki} = \left[\sum_{j=1}^{c} \left(\frac{D(x(t), m_i(t))}{D(x(t), m_j(t))} \right)^{\frac{1}{m-1}} \right]^{-1}, \tag{2.68}$$

for $i = 1, 2, \dots, c$.
FLC5. Update $m_1(t), \dots, m_c(t)$:

$$m_l(t+1) = m_l(t) + \alpha(t) H(u_{kl})[x(t) - m_l(t)], \tag{2.69}$$

for $l = 1, 2, \dots, c$.
End FLC.

In (2.69), $H\colon [0,1] \to [0,1]$ is either linear or a sigmoid function such that $H(0) = 0$ and $H(1) = 1$.

There are variations of **FLC**. For example, the step FLC5 can be replaced by FLC5': Let $\ell = \arg \max_{1 \le j \le c} u_{kj}$. Then

$$m_\ell(t+1) = m_\ell(t) + \alpha(t) H(u_{k\ell})[x(t) - m_\ell(t)],$$
$$m_i(t+1) = m_i(t), \ i \neq \ell.$$

As many other variations of the competitive learning algorithms have been proposed, the corresponding fuzzy learning methods can be derived without difficulty. Fundamentally the fuzzy learning algorithms are heuristic and not based on a rigorous mathematical theory, however.

2.11 Direct Derivation of Classification Functions

Fuzzy classification rules (2.53) and (2.54) have been derived from the membership matrices by replacing an object by the variable x. This derivation is somewhat artificial and does not uncover a fundamental idea behind the rules.

A fuzzy classification rule is important in itself apart from fuzzy c-means. Hence direct derivation of a fuzzy classification rule without the concept of clustering should be considered.

To this end, we first notice that the classification rules should be determined using prototypes $v_1, \dots, v_c$. In clustering these are cluster centers but they are not necessarily centers but may be other prototypes in the case of supervised classification.

Assume $v_1, \ldots, v_c$ are given and suppose we wish to determine a classification rule of nearest prototype. The solution is evident:

$$\mathrm{U}_1^{(i)}(x) = \begin{cases} 1 & (v_i = \arg\min_{1 \le j \le c} D(x, v_j)), \\ 0 & (\text{otherwise}). \end{cases} \tag{2.70}$$

For technical reason we employ a closed ball $B(r)$ with the radius r:

$$B(r) = \{\ x \in \boldsymbol{R}^p\ :\ \|x\| \le r\ \} \tag{2.71}$$

where r is sufficiently large so that it contains all prototypes $v_1, \ldots, v_c$, and we consider the problem inside this region.

Note the above function is the optimal solution of the following problem:

$$\min_{\mathrm{U}_j, j=1,\ldots,c} \sum_{j=1}^{c} \int_{B(r)} \mathrm{U}_j(x) D(x, v_j) dx \tag{2.72}$$

$$\text{subject to } \sum_{j=1}^{c} \mathrm{U}_j(x) = 1, \quad \mathrm{U}_\ell(x) \ge 0, \quad \ell = 1, \ldots, c. \tag{2.73}$$

We fuzzify this function. We note the above function is not differentiable. We 'regularize' the function by considering a differentiable approximation of this function.

For this purpose we add an entropy term and consider the following.

$$\min_{\mathrm{U}_j, j=1,\ldots,c} \sum_{j=1}^{c} \int_{B(r)} \mathrm{U}_j(x) D(x, v_j) dx + \nu \sum_{j=1}^{c} \int_{B(r)} \mathrm{U}_j(x) \log \mathrm{U}_j(x) dx \tag{2.74}$$

$$\text{subject to } \sum_{j=1}^{c} \mathrm{U}_j(x) = 1, \quad \mathrm{U}_\ell(x) \ge 0,\ \ell = 1, \ldots, c. \tag{2.75}$$

To obtain the optimal solution we employ the calculus of variations. Let

$$J = \sum_{j=1}^{c} \int_{B(r)} \mathrm{U}_j(x) D(x, v_j) dx + \nu \sum_{j=1}^{c} \int_{B(r)} \mathrm{U}_j(x) \log \mathrm{U}_j(x) dx$$

and notice the constraint. Hence we should minimize the Lagrangian

$$L = J + \int_{B(r)} \lambda(x) [\sum_{j=1}^{c} \mathrm{U}_j(x) - 1] dx.$$

Put $\mathbf{U}(x) = (\mathrm{U}_1(x), \ldots, \mathrm{U}_c(x))$ for simplicity. Let

$$\epsilon \delta L + o(\epsilon^2) = L(\mathbf{U} + \epsilon \eta_i, \lambda) - L(\mathbf{U}, \lambda).$$

where

$$[\mathbf{U} + \epsilon \eta_i](x) = (\mathrm{U}_1(x), \ldots, \mathrm{U}_{i-1}(x), \mathrm{U}_i(x) + \epsilon \eta_i(x), \mathrm{U}_{i+1}(x), \ldots, \mathrm{U}_c(x)).$$

We then have

$$\delta L = \int_{B(r)} \eta_i(x) D(x, v_i) dx + \nu \int_{B(r)} \eta_i(x)[1 + \log \mathrm{U}_i(x)] dx$$
$$+ \int_{B(r)} \eta_i(x) \lambda(x) dx.$$

Put $\delta L = 0$ and note $\eta_i(x)$ is arbitrary. We hence have

$$D(x, v_i) + \nu(1 + \log \mathrm{U}_i(x)]) + \lambda(x) = 0$$

from which

$$\mathrm{U}_i(x) = \exp(-1 - \lambda(x)/\nu) \exp(-D(x, v_i)/\nu)$$

holds. Summing up the above equation with respect to $j = 1, \ldots, c$:

$$1 = \sum_{j=1}^{c} \mathrm{U}_j(x) = \sum_{j=1}^{c} \exp(-1 - \lambda(x)/\nu) \exp(-D(x, v_j)/\nu).$$

We thus obtain the solution

$$\mathrm{U}_i(x) = \frac{\exp(-D(x, v_i)/\nu)}{\sum_{j=1}^{c} \exp(-D(x, v_j)/\nu)}, \tag{2.76}$$

which is the same as (2.54).

For the classification function (2.53), the same type of the calculus of variations should be applied. Thus the problem to be solved is

$$\min_{\mathrm{U}_j, j=1,\ldots,c} \sum_{j=1}^{c} \int_{B(r)} (\mathrm{U}_j(x))^m D(x, v_j) dx \tag{2.77}$$

$$\text{subject to} \sum_{j=1}^{c} \mathrm{U}_j(x) = 1, \quad \mathrm{U}_\ell(x) \geq 0, \quad \ell = 1, \ldots, c. \tag{2.78}$$

The optimal solution is, as we expect,

$$\mathrm{U}_i(x) = \frac{1/D(x, v_i)^{\frac{1}{m-1}}}{\sum_{j=1}^{c} 1/D(x, v_j)^{\frac{1}{m-1}}}. \tag{2.79}$$

We omit the derivation, as it is similar to the above. As the last remark, we note that the radius r can be arbitrarily large in $B(r)$. Therefore the classification function can be assumed to be defined on the space $\boldsymbol{R}^p$.

The formulation by the calculus of variations in this section thus justifies the use of these functions in the fuzzy learning and the fixed point iterations.

2.12 Mixture Density Model and the EM Algorithm

Although this book mainly discusses fuzzy clustering, a statistical model is closely related to the methods of fuzzy c-means. In this section we overview the mixture density model that is frequently employed for both supervised and unsupervised classification [25, 98, 131]. For this purpose we use terms in probability and statistics in this section.

Although probability density functions for most standard distributions are unimodal, we note that clustering should handle multimodal distributions.

Let us for the moment suppose that many data are collected and the histogram has two modes of maximal values. Apparently, the histogram is represented by mixing two densities of unimodal distributions. Hence the probability density is

$$p(x) = \alpha_1 p_1(x) + \alpha_2 p_2(x),$$

where α_1 and α_2 are nonnegative numbers such that $\alpha_1 + \alpha_2 = 1$. Typically, the two distributions are normal:

$$p_i(x) = \frac{1}{\sqrt{2\pi}\sigma_i} e^{-\frac{(x-\mu_i)^2}{2\sigma_i{}^2}}, \qquad i = 1, 2.$$

Suppose moreover that we have good estimates of the parameters α_i, μ_i, and σ_i, $i = 1, 2$. After having a good approximation of the mixture distribution we can solve the clustering problem using the Bayes formula for posterior probability as follows.

Let us assume $P(X|C_i)$ and $P(C_i)$ $(i = 1, \dots, m)$ be the conditional probability of event X given that class C_i occurs, and the prior probability of the class C_i, respectively. We assume that exactly one of C_i, $i = 1, \dots, m$, necessarily occurs. The Bayes formula is used for determining the class of X:

$$P(C_i|X) = \frac{P(X|C_i)P(C_i)}{\sum_{j=1}^{m} P(X|C_j)P(C_j)}. \qquad (2.80)$$

Let us apply the formula (2.80) to the above example: put

$$P(X) = P(a < x < b), \quad P(C_i) = \alpha_i, \quad P(X|C_i) = \int_a^b p_i(x)dx \quad (i = 1, 2).$$

Then we have

$$P(C_i|X) = \frac{\alpha_i \int_a^b p_i(x)dx}{\sum_{j=1}^{2} \alpha_j \int_a^b p_j(x)dx}.$$

Assume we have observation y. Taking two numbers a, b such that $a < y < b$, we have the probability of the class C_i given X by the above formula. Take the limit $a \to y$ and $b \to y$, we then have the probability of the class C_i given the data y:

$$P(C_i|y) = \frac{\alpha_i p_i(y)}{\sum_{j=1}^{2} \alpha_j p_j(y)}. \tag{2.81}$$

As this gives us the probability of allocating an observation to each class, the clustering problem is solved; instead of fuzzy membership, we have the probability of membership to a class.

Note the above formula (2.81) is immediately generalized to the case of m classes.

The problem is how to obtain good estimates of the parameters. The EM algorithm should be used for this purpose.

2.12.1 The EM Algorithm

Consider a general class of mixture distribution given by

$$p(x|\Phi) = \sum_{j=1}^{m} \alpha_i p_i(x|\phi_i) \tag{2.82}$$

in which $p_i(x|\phi_i)$ is the probability density corresponding to class C_i, and ϕ_i is a vector parameter to be estimated. Moreover Φ represents the whole sequence of the parameters, i.e., $\Phi = (\alpha_1, \dots, \alpha_m, \phi_1, \dots, \phi_m)$. We assume that observation $x_1, \dots, x_n$ are mutually independent samples taken from the population having this mixture distribution. The symbols $x_1, \dots, x_n$ are used for both observation and variables for the sample distribution. Although this is an abuse of terminology for simplicity, no confusion arises.

A classical method to solve a parameter estimation problem is the maximum likelihood. From the assumption of independence, the sample distribution is given by

$$\prod_{k=1}^{N} p(x_k|\Phi).$$

Suppose x_k is the observation, then the above is a function of parameter Φ. The maximum likelihood is the method to use the parameter value that maximizes the above *likelihood function*. For convenience in calculations, the *log-likelihood* is generally used:

$$L(\Phi) = \log \prod_{k=1}^{N} p(x_k|\Phi) = \sum_{k=1}^{N} \log p(x_k|\Phi). \tag{2.83}$$

Thus the maximum likelihood estimate is given by

$$\hat{\Phi} = \arg\max_{\Phi} L(\Phi). \tag{2.84}$$

For simple distributions, the maximum likelihood estimates are easy to calculate, but an advanced method should be used for this mixture distribution.

The EM algorithm [25, 98, 131] is useful for this purpose. The EM algorithm is an iterative procedure in which an Expectation step (E-step) and a Maximization step (M-step) are repeated until convergence.

1. In addition to the observation $x_1, \dots, x_n$, assume that $y_1, \dots, y_n$ represents *complete data*. In contrast, $x_1, \dots, x_n$ is called *incomplete data*. For simplicity, we write $\boldsymbol{x} = (x_1, \dots, x_n)$ and $\boldsymbol{y} = (y_1, \dots, y_n)$. Actually, $\boldsymbol{y}$ itself is not observed; only partial observation of the incomplete data $\boldsymbol{x}$ is available. Let us assume the mapping from the complete data to the corresponding incomplete data be $\chi \colon \boldsymbol{y} \mapsto \boldsymbol{x}$. Given $\boldsymbol{x}$, the set of all $\boldsymbol{y}$ such that $\boldsymbol{x} = \chi(\boldsymbol{y})$ is thus given by the inverse image $\chi^{-1}(\boldsymbol{x})$.
2. We assume that $\boldsymbol{x}$ and $\boldsymbol{y}$ have probability functions $g(\boldsymbol{x}|\Phi)$ and $f(\boldsymbol{y}|\Phi)$, respectively.
3. Assume that an estimate Φ' for Φ is given. The soul of the EM algorithm is to optimize the next function $Q(\Phi|\Phi')$:

$$Q(\Phi|\Phi') = E(\log f(\boldsymbol{y}|\Phi)|\boldsymbol{x}, \Phi') \tag{2.85}$$

   where $E(\log f|\boldsymbol{x}, \Phi')$ is the conditional expectation given $\boldsymbol{x}$ and Φ'.

Let us assume that $k(\boldsymbol{y}|\boldsymbol{x}, \Phi')$ is the conditional probability function of $\boldsymbol{y}$ given $\boldsymbol{x}$ and Φ'. It then follows that

$$Q(\Phi|\Phi') = \sum_{\boldsymbol{y} \in \chi^{-1}(\boldsymbol{x})} k(\boldsymbol{y}|\boldsymbol{x}, \Phi') \log f(\boldsymbol{y}|\Phi). \tag{2.86}$$

We are now ready to describe the EM algorithm.

The EM algorithm

(O) Set an initial estimate $\Phi^{(0)}$ for Φ. Let $\ell = 0$. Repeat the following (E) and (M) until convergence.

(E) (Expectation Step)
Calculate $Q(\Phi|\Phi^{(\ell)})$.

(M) (Maximization Step)
Find the maximizing solution

$$\bar{\Phi} = \arg\max_{\Phi} Q(\Phi|\Phi^{(\ell)}).$$

Let $\ell := \ell + 1$ and $\Phi^{(\ell)} = \bar{\Phi}$.

Note that the (E) and (M) steps are represented by a single formula:

$$\Phi^{(\ell+1)} = \arg\max_{\Phi} Q(\Phi|\Phi^{(\ell)}), \qquad \ell = 1, 2, \dots$$

for $\ell = 0, 1, \dots$ until convergence.

2.12.2 Parameter Estimation in the Mixture Densities

The EM algorithm is applied to the present class of the mixture distributions. For this purpose what the complete data in this case should be clarified. Suppose we have the information, in addition to x_k, from which class C_i the observation has been obtained. Then the estimation problem becomes simpler using this information. Hence we assume $y_k = (x_k, i_k)$ in which i_k means the class number of C_{i_k} from which x_k has been obtained.

Given this information, the density $f(\boldsymbol{y}|\Phi)$ is

$$f(\boldsymbol{y}|\Phi) = \prod_{k=1}^{N} \alpha_{i_k} p_{i_k}(x_k|\phi_{i_k}).$$

The conditional density $k(\boldsymbol{y}|\boldsymbol{x}, \Phi')$ is then calculated as follows.

$$k(\boldsymbol{y}|\boldsymbol{x}, \Phi') = \frac{f(\boldsymbol{y}|\Phi')}{g(\boldsymbol{x}|\Phi')} = \prod_{k=1}^{N} \frac{\alpha'_{i_k} p_{i_k}(x_k|\phi'_{i_k})}{p(x_k|\Phi')}.$$

Notice that $g(\boldsymbol{x}|\Phi') = \prod_{k=1}^{N} p(x_k|\Phi')$.

We now can calculate $Q(\Phi|\Phi')$. It should be noted that $\chi^{-1}(\boldsymbol{x})$ is reduced to the set $\{(i_1, \ldots, i_n) : 1 \le i_\ell \le m,\ \ell = 1, \ldots, m\}$. We have

$$Q(\Phi|\Phi') = \sum_{i_1=1}^{m} \cdots \sum_{i_n=1}^{m} \sum_{k=1}^{N} \log[\alpha_{i_k} p_{i_k}(x_k|\phi_{i_k})] \prod_{k=1}^{N} \frac{\alpha'_{i_k} p_{i_k}(x_k|\phi'_{i_k})}{p(x_k|\Phi')}.$$

After straightforward calculations, we have

$$Q(\Phi|\Phi') = \sum_{i=1}^{m} \sum_{k=1}^{N} \log[\alpha_i p_i(x_k|\phi_i)] \frac{\alpha'_i p_i(x_k|\phi'_i)}{p(x_k|\Phi')}.$$

For simplicity, put

$$\psi_{ik} = \frac{\alpha'_i p_i(x_k|\phi'_i)}{p(x_k|\Phi')}, \quad \Psi_i = \sum_{k=1}^{N} \psi_{ik}.$$

and note that $\sum_{i=1}^{m} \Psi_i = n$. It then follows that

$$Q(\Phi|\Phi') = \sum_{i=1}^{m} \Psi_i \log \alpha_i + \sum_{i=1}^{m} \sum_{k=1}^{N} \psi_{ik} \log p_i(x_k|\phi_i).$$

To obtain the optimal α_i, we should take the constraint $\sum_{i=1}^{m} \alpha_i = 1$ into account. Hence the Lagrangian with the multiplier λ is used:

$$L = Q(\Phi|\Phi') - \lambda(\sum_{i=1}^{m} \alpha_i - 1)$$

Using

$$\frac{\partial L}{\partial \alpha_i} = \frac{\Psi_i}{\alpha_i} - \lambda = 0, \tag{2.87}$$

and taking the sum of $\lambda\alpha_i = \Psi_i$ with respect to $i = 1, \ldots, m$, we have $\lambda = N$. Thus, the optimal solution is

$$\alpha_i = \frac{\Psi_i}{N} = \frac{1}{N}\sum_{i=1}^{N} \frac{\alpha_i' p_i(x_k|\phi_i')}{p(x_k|\Phi')}, \qquad i = 1, \ldots, m. \tag{2.88}$$

Normal distributions

We have not specified the density functions until now. We proceed to consider normal distributions and estimate the means and variances. For simplicity we first derive solutions for the univariate distributions. After that we show the solutions for multivariate normal distributions.

For the univariate normal distributions,

$$p_i(x|\phi_i) = \frac{1}{\sqrt{2\pi}\sigma_i} e^{-\frac{(x-\mu_i)^2}{2\sigma_i{}^2}}, \quad i = 1, \ldots, m$$

where $\phi_i = (\mu_i, \sigma_i)$. For the optimal solution we should minimize

$$J = \sum_{i=1}^{m}\sum_{k=1}^{N} \psi_{ik} \log \frac{1}{\sqrt{2\pi}\sigma_i} e^{-\frac{(x_k-\mu_i)^2}{2\sigma_i{}^2}}.$$

From

$$\frac{\partial J}{\partial \mu_i} = -\sum_{k=1}^{N} \psi_{ik} \frac{x_k - \mu_i}{\sigma_i^2} = 0,$$

we have

$$\mu_i = \frac{1}{\Psi_i}\sum_{k=1}^{N} \psi_{ik} x_k, \qquad i = 1, \ldots, m. \tag{2.89}$$

In the same manner, from

$$\frac{\partial J}{\partial \sigma_i} = \sum_{k=1}^{N} \psi_{ik} \frac{(x_k - \mu_i)^2}{\sigma_i^3} - \sum_{k=1}^{N} \psi_{ik} \frac{1}{\sigma_i} = 0,$$

We have

$$\sigma_i^2 = \frac{1}{\Psi_i}\sum_{k=1}^{N} \psi_{ik}(x_k - \mu_i)^2 = \frac{1}{\Psi_i}\sum_{k=1}^{N} \psi_{ik} x_k^2 - \mu_i^2, \quad i = 1, \ldots, m, \tag{2.90}$$

in which μ_i is given by (2.89).

Let us next consider multivariate normal distributions:

$$p_i(x) = \frac{1}{2\pi^{\frac{p}{2}} |\Sigma_i|^{\frac{1}{2}}} e^{-\frac{1}{2}(x-\mu_i)^\top \Sigma_i{}^{-1}(x-\mu_i)}$$

in which $x = (x^1, \ldots, x^p)$ and $\mu_i = (\mu_i^1, \ldots, \mu_i^p)$ are vectors, and $\Sigma_i = (\sigma_i^{j\ell})$ $(1 \le j, \ell \le p)$ is the covariance matrix; $|\Sigma_i|$ is the determinant of Σ_i.

By the same manner as above, the optimal α_i is given by (2.88), while the solutions for μ_i and Σ_i are as follows [131].

$$\mu_i = \frac{1}{\Psi_i} \sum_{k=1}^{N} \psi_{ik} x_k, \qquad i = 1, \ldots, m, \tag{2.91}$$

$$\Sigma_i = \frac{1}{\Psi_i} \sum_{k=1}^{N} \psi_{ik} (x_k - \mu_i)(x_k - \mu_i)^\top, \qquad i = 1, \ldots, m. \tag{2.92}$$

Proofs of formulas for multivariate normal distributions

Let us prove (2.91) and (2.92). Readers who are uninterested in mathematical details may skip the proof.

Let jth component of vector μ_i be μ_i^j or $(\mu_i)^j$, and (i, ℓ) component of matrix Σ_i be $\sigma_i^{j\ell}$ or $(\Sigma_i)^{j\ell}$. A matrix of which (i, j) component is f^{ij} is denoted by $[f^{ij}]$. Thus, $\mu_i^j = (\mu_i)^j$ and $\Sigma_i = [\sigma_i^{j\ell}] = [(\Sigma_i)^{j\ell}]$.

Let

$$J_i = \sum_{k=1}^{N} \psi_{ik} \log \frac{1}{2\pi^{\frac{p}{2}} |\Sigma_i|^{\frac{1}{2}}} e^{-\frac{1}{2}(x_k - \mu_i)^\top \Sigma_i{}^{-1}(x_k - \mu_i)}$$

and notice that we should find the solutions of $\frac{\partial J_i}{\partial \mu_i^j} = 0$ and $\frac{\partial J_i}{\partial \sigma_i^{j\ell}} = 0$. It is immediate to see that the solution of $\frac{\partial J_i}{\partial \mu_i^j} = 0$ is given by (2.91), and hence the detail is omitted.

Let us consider solution of $\frac{\partial J_i}{\partial \sigma_i^{j\ell}} = 0$, i.e.,

$$\begin{aligned} 2\frac{\partial J_i}{\partial \sigma_i^{j\ell}} = &-\sum_{k=1}^{N} \psi_{ik} \frac{\partial}{\partial \sigma_i^{j\ell}} \left((x_k - \mu_i)^\top \Sigma_i (x_k - \mu_i) \right) \\ &- \frac{\partial}{\partial \sigma_i^{j\ell}} (\log |\Sigma_i|) = 0 \end{aligned}$$

To solve this, we note the following.

(i) Let the cofactor for $\sigma_i^{j\ell}$ in the matrix Σ_i be $\tilde{\Sigma}_i^{j\ell}$. We then have

$$\left[\frac{\partial}{\partial \sigma_i^{j\ell}} \log |\Sigma_i| \right] = \left[\frac{1}{|\Sigma_i|} \frac{\partial}{\partial \sigma_i^{j\ell}} |\Sigma_i| \right] = \left[\frac{1}{|\Sigma_i|} \tilde{\Sigma}_i^{j\ell} \right] = \Sigma^{-1}$$

(ii) To calculate $\frac{\partial}{\partial \sigma_i^{j\ell}} \Sigma_i^{-1}$, let $E_{j\ell}$ is the matrix in which the (j, ℓ) component alone is the unity and all other components are zero. That is,

$$(E_{j\ell})^{ik} = \delta_{ij} \delta_{k\ell}$$

using the Kronecker delta δ_{ij}. Then,

$$\frac{\partial}{\partial \sigma_i^{j\ell}}(\Sigma_i^{-1}\Sigma_i) = \left(\frac{\partial}{\partial \sigma_i^{j\ell}}\Sigma_i^{-1}\right)\Sigma_i + \Sigma_i^{-1}\left(\frac{\partial}{\partial \sigma_i^{j\ell}}\Sigma_i\right)$$
$$= \left(\frac{\partial}{\partial \sigma_i^{j\ell}}\Sigma_i^{-1}\right)\Sigma_i + \Sigma_i E_{j\ell} = 0$$

whereby we have

$$\frac{\partial}{\partial \sigma_i^{j\ell}}\Sigma_i^{-1} = -\Sigma_i^{-1}E_{j\ell}\Sigma_i^{-1}.$$

Suppose a vector ξ does not contain an element in Σ_i. We then obtain

$$\left[\frac{\partial}{\partial \sigma_i^{j\ell}}(\xi^\top \Sigma_i^{-1}\xi)\right] = -\left[\xi^\top \Sigma_i^{-1}E_{j\ell}\Sigma_i^{-1}\xi\right]$$
$$= -\left[(\Sigma_i^{-1}\xi)^\top E_{j\ell}(\Sigma_i^{-1}\xi)\right]$$
$$= -\left[(\Sigma_i^{-1}\xi)^j(\Sigma_i^{-1}\xi)^\ell\right]$$
$$= -(\Sigma_i^{-1}\xi)(\Sigma_i^{-1}\xi)^\top = -\Sigma_i^{-1}\xi\xi^\top\Sigma_i^{-1}.$$

Using (i) and (ii) in the last equation, we obtain

$$2\frac{\partial J_i}{\partial \sigma_i^{j\ell}} = \sum_{k=1}^{N}\psi_{ik}\Sigma_i^{-1}(x_k-\mu_i)(x_k-\mu_i)^\top\Sigma_i^{-1} - \left(\sum_{k=1}^{N}\psi_{ik}\right)\Sigma_i^{-1} = 0.$$

Multiplications of Σ_i to the above equation from the right and the left lead us to

$$-\sum_{k=1}^{N}\psi_{ik}(x_k-\mu_i)(x_k-\mu_i)^\top + \Psi_i\Sigma_i = 0.$$

Thus we obtain (2.92).

3 Variations and Generalizations - I

Many studies have been done with respect to variations and generalizations of the basic methods of fuzzy c-means. We will divide those variations and generalizations into two classes. The first class has 'standard variations or generalizations' that include relatively old studies, or should be known to many readers of general interest. On the other hand, the second class includes more specific studies or those techniques for a limited purpose and will be interested in by more professional readers. We describe some algorithms in the first class in this chapter.

3.1 Possibilistic Clustering

Krishnapuram and Keller [87] propose the method of possibilistic clustering: the same alternate optimization algorithm **FCM** is used in which the constraint $\mathcal{U}_f$ is not employed but nontrivial solution of

$$\sum_{k=1}^{N} u_{ki} > 0, \quad 1 \le i \le c; \quad u_{kj} \ge 0, \quad 1 \le k \le n, \; 1 \le j \le c \tag{3.1}$$

should be obtained.

For this purpose the objective function J_{fcm} cannot be used since the optimal $\bar{U}$ is trivial: $\bar{u}_{ki} = 0$ for all i and k. Hence a modified objective function

$$J_{\text{pos}}(U, V) = \sum_{i=1}^{c} \sum_{k=1}^{N} (u_{ki})^m D(x_k, v_i) + \sum_{i=1}^{c} \eta_i \sum_{k=1}^{N} (1 - u_{ki})^m \tag{3.2}$$

has been proposed. The solution $\bar{U}$ becomes

$$\bar{u}_{ki} = \frac{1}{1 + \left(\frac{D(x_k, \bar{v}_i)}{\eta_i} \right)^{\frac{1}{m-1}}} \tag{3.3}$$

S. Miyamoto et al.: Algorithms for Fuzzy Clustering, STUDFUZZ 229, pp. 43–66, 2008.
springerlink.com

while the optimal $\bar{v}_i$ remains the same:

$$\bar{v}_i = \frac{\sum_{k=1}^{N} (\bar{u}_{ki})^m x_k}{\sum_{k=1}^{N} (\bar{u}_{ki})^m}.$$

Notice that the fuzzy classification function derived from the above $\bar{U}$ is

$$\mathbf{U}_{\text{pos}}(x; v_i) = \frac{1}{1 + \left(\frac{D(x, v_i)}{\eta_i}\right)^{\frac{1}{m-1}}} \tag{3.4}$$

We will observe other types of possibilistic clustering in which we obtain different classification functions. A classification function from a method of possibilistic clustering in general is denoted by $\mathbf{U}(x; v_i)$. Notice that this form is different from that for fuzzy c-means: the latter is $\mathbf{U}^{(i)}(x; V)$ with the superscript (i) and the parameter V, while the former is without the superscript and the parameter is just v_i.

The classification function $\mathbf{U}_{\text{pos}}(x; v_i)$ has the next properties, when we put $\mathbf{U}(x; v_i) = \mathbf{U}_{\text{pos}}(x; v_i)$.

(i) $\mathbf{U}(x; v_i)$ is unimodal with the maximum value at $x = v_i$.

(ii)

$$\max_{x \in \boldsymbol{R}^p} \mathbf{U}(x; v_i) = \mathbf{U}(v_i; v_i) = 1.$$

(iii)

$$\inf_{x \in \boldsymbol{R}^p} \mathbf{U}(x; v_i) = 0.$$

(iv) Let us define the set $X_i \subset \boldsymbol{R}^p$ by

$$X_i = \{\, x \in \boldsymbol{R}^p \;:\; \mathbf{U}(x; v_i) > \mathbf{U}(x; v_j),\ \forall j \neq i \,\}.$$

Then, $X_1, \ldots, X_c$ are the Voronoi sets of $\boldsymbol{R}^p$.

3.1.1 Entropy-Based Possibilistic Clustering

The above objective function should not be the only one for the possibilistic clustering [23]; the entropy-based objective function (6.4) can also be used for the possibilistic method:

$$J_{\text{efc}}(U, V) = \sum_{i=1}^{c} \sum_{k=1}^{N} u_{ki} D(x_k, v_i) + \nu \sum_{i=1}^{c} \sum_{k=1}^{N} u_{ki} \log u_{ki}.$$

The solution $\bar{U}$ and $\bar{V}$ are respectively given by

$$\bar{u}_{ki} = \exp\left(-1 - \frac{D(x_k, \bar{v}_i)}{\nu}\right) \tag{3.5}$$

and

$$\bar{v}_i = \frac{\sum_{k=1}^{N} \bar{u}_{ki} x_k}{\sum_{k=1}^{N} \bar{u}_{ki}}.$$

The corresponding classification function is $\mathbf{U}'(x; v_i) = \exp\left(-1 - \frac{D(x,v_i)}{\nu}\right)$, but this function should be modified, since $\mathbf{U}'(x; v_i)$ satisfies the above (i), (iii), (iv), but not (ii). Apparently

$$\bar{u}_{ki} = \exp\left(-\frac{D(x_k, \bar{v}_i)}{\nu}\right) \tag{3.6}$$

and

$$\mathbf{U}_{\text{pose}}(x; v_i) = \exp\left(-\frac{D(x, v_i)}{\nu}\right) \tag{3.7}$$

are simpler and $\mathbf{U}_{\text{pose}}(x; v_i)$ satisfies (i–iv). Thus, instead of (6.4),

$$J_{\text{pose}}(U, V) = \sum_{i=1}^{c} \sum_{k=1}^{N} u_{ki} D(x_k, v_i) + \nu \sum_{i=1}^{c} \sum_{k=1}^{N} u_{ki}(\log u_{ki} - 1). \tag{3.8}$$

is used, whereby we have the solution (3.6).

Note 3.1.1. We can derive the classification functions of possibilistic clustering using calculus of variations as in Section 2.11. We hence consider

$$\min_{\mathrm{U}_j,\ j=1,\ldots,c} \sum_{j=1}^{c} \int_{\boldsymbol{R}^p} (\mathrm{U}_j(x))^m D(x, v_j) dx + \sum_{j=1}^{c} \eta_j \int_{\boldsymbol{R}^p} (1 - \mathrm{U}_j(x))^m dx \tag{3.9}$$

or

$$\min_{\mathrm{U}_j,\ j=1,\ldots,c} \sum_{j=1}^{c} \int_{\boldsymbol{R}^p} \mathrm{U}_j(x) D(x, v_j) dx + \nu \sum_{j=1}^{c} \int_{\boldsymbol{R}^p} \mathrm{U}_j(x) \log(\mathrm{U}_j(x) - 1) dx \tag{3.10}$$

The method is the same with the exception that the constraint $\sum_{i=1}^{c} u_{ki} = 1$ is unnecessary. We omit the detail.

Note 3.1.2. Observe that $\mathbf{U}_{\text{pose}}(x; v_i)$ approaches zero more rapidly than $\mathbf{U}_{\text{pos}}(x; v_i)$ when $\|x\| \to \infty$, since it is easy to see that

$$\frac{\mathbf{U}_{\text{pos}}(x; v_i)}{\mathbf{U}_{\text{pose}}(x; v_i)} \to \infty, \quad \text{as } \|x\| \to \infty$$

from the Taylor expansion of the exponential function.

3.1.2 Possibilistic Clustering Using a Quadratic Term

The objective function (2.42) can also be employed for possibilistic clustering [110]:

$$J_{\text{qfc}}(U,V) = \sum_{i=1}^{c}\sum_{k=1}^{N} u_{ki} D(x_k, v_i) + \frac{1}{2}\nu \sum_{i=1}^{c}\sum_{k=1}^{N} u_{ki}^2 \quad (\nu > 0), \tag{3.11}$$

where we should note the obvious constraint $u_{ki} \geq 0$. The same technique to put $w_{ki}^2 = u_{ki}$ is useful and we derive the solution

$$\bar{u}_{ki} = \begin{cases} 1 - \frac{D(x_k,\bar{v}_i)}{\nu} & (\frac{D(x_k,\bar{v}_i)}{\nu} < 1), \\ 0 & (\text{otherwise}). \end{cases} \tag{3.12}$$

with the same formula for the cluster centers: $\bar{v}_i = \sum_{k=1}^{N} \bar{u}_{ki} x_k \Big/ \sum_{k=1}^{N} \bar{u}_{ki}$. The classification function is

$$\mathbf{U}_{\text{posq}}(x; v_i) = \begin{cases} 1 - \frac{D(x,\bar{v}_i)}{\nu} & (\frac{D(x,\bar{v}_i)}{\nu} < 1), \\ 0 & (\text{otherwise}). \end{cases} \tag{3.13}$$

3.1.3 Objective Function for Fuzzy c-Means and Possibilistic Clustering

We observe the objective function (2.20) of the standard fuzzy c-means cannot be used for possibilistic clustering and hence another objective function (3.2) has been proposed. In contrast, the entropy-based objective function (3.8) and (3.11) can be used for both fuzzy c-means and possibilistic clustering. Notice $J_{\text{pose}}(U,V) = J_{\text{efc}}(U,V) - N$ with the constraint $\sum_{i=1}^{c} u_{ki} = 1$.

A question naturally arises whether or not it is possible to use the same objective function of the standard fuzzy c-means or possibilistic clustering for the both methods of clustering. We have two answers for this question.

First, notice that the function $F_{DB}(x; v_i) = \frac{1}{D(x,v_i)^{\frac{1}{m-1}}}$. cannot be employed as the classification function for possibilistic clustering, since this function has the singularity at $x = v_i$. The converse is, however, possible: we can use

$$\mathbf{U}^{(i)}_{\text{posfcm}}(x; V) = \frac{\mathbf{U}_{\text{pos}}(x; v_i)}{\sum_{j=1}^{c} \mathbf{U}_{\text{pos}}(x; v_j)}$$

$$\mathbf{U}_{\text{pos}}(x; v_i) = \frac{1}{1 + \left(\frac{D(x,v_i)}{\eta_i}\right)^{\frac{1}{m-1}}}$$

as the classification function for the membership of fuzzy c-means. In other words, we repeat

$$u_{ki} = \mathbf{U}^{(i)}_{\text{posfcm}}(x_k; V);$$

$$v_i = \frac{\sum_{k=1}^{N} (u_{ki})^m x_k}{\sum_{k=1}^{N} (u_{ki})^m};$$

until convergence as an algorithm of fixed point iteration.

Second solution is based on a rigorous alternate optimization [147]. Instead, the objective function should be limited. We consider

$$J_{\text{pos}}(U, V) = \sum_{i=1}^{c} \sum_{k=1}^{N} (u_{ki})^2 D(x_k, v_i) + \sum_{i=1}^{c} \eta_i \sum_{k=1}^{N} (1 - u_{ki})^2 \tag{3.14}$$

which is the same as (3.2) except that $m = 2$ is assumed. The solution u_{ik} for possibilistic clustering is obvious:

$$\bar{u}_{ki} = \frac{1}{1 + \frac{D(x_k, \bar{v}_i)}{\eta_i}}.$$

For the fuzzy c-means solution with the constraint $\sum_{i=1}^{c} u_{ki} = 1$, we have

$$u_{ki} = \frac{1 + \frac{D(x_k, \bar{v}_i)}{\eta_i}}{\sum_{j=1}^{c} \left(1 + \frac{D(x_k, \bar{v}_j)}{\eta_j}\right)}.$$

Define

$$F_{DB2}(x; v_i) = \frac{1}{1 + \frac{D(x, v_i)}{\eta_i}}.$$

We then have the two classification functions of this method:

$$\mathbf{U}_{\text{pos}}(x; v_i) = F_{DB2}(x; v_i),$$

$$\mathbf{U}^{(i)}_{\text{posfcm}}(x; V) = \frac{F_{DB2}(x; v_i)}{\sum_{j=1}^{c} F_{DB2}(x; v_j)}.$$

3.2 Variables for Controlling Cluster Sizes

As seen in the previous chapter, the obtained clusters are in the Voronoi sets when the crisp reallocation rule is applied. Notice that the Voronoi sets have piecewise linear boundaries and the reallocation uses the nearest center rule. This means that an algorithm of fuzzy c-means or crisp c-means may fail to divide even well-separated groups of objects into clusters.

Let us observe an example of objects in Figure 3.1 in which a group of 35 objects and another of 135 objects are seen. Figure 3.2 shows the result of clustering

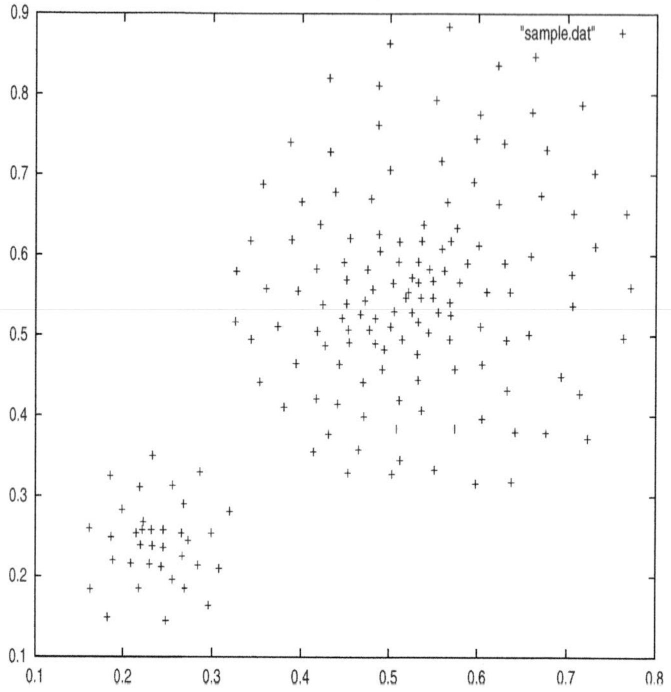

Fig. 3.1. An artificially generated data set which has two groups of 35 and 135 objects

using an algorithm of the crisp c-means. The same result has been obtained by using the fuzzy c-means and crisp reallocation by the maximum membership rule. Readers can observe a part of the larger group is judged to be in the smaller cluster. This result is not strange in view of the above rule of the nearest center. Suppose the two cluster centers are near the central parts of the two groups. Then the linear line equidistant from the two centers crosses the larger group, and hence some objects in the larger group is classified into the smaller cluster.

Such misclassifications arise in many real applications. Therefore a new method is necessary to overcome such a difficulty.

A natural idea is to use an additional variable that controls cluster sizes, in other words, cluster volumes. We consider the following two objective functions for this purpose [68, 111].

$$J_{\text{efca}}(U, V, A) = \sum_{i=1}^{c} \sum_{k=1}^{N} u_{ki} D(x_k, v_i) + \nu \sum_{i=1}^{c} \sum_{k=1}^{N} u_{ki} \log \frac{u_{ki}}{\alpha_i}, \tag{3.15}$$

$$J_{\text{fcma}}(U, V, A) = \sum_{i=1}^{c} \sum_{k=1}^{N} (\alpha_i)^{1-m} (u_{ki})^m D(x_k, v_i). \tag{3.16}$$

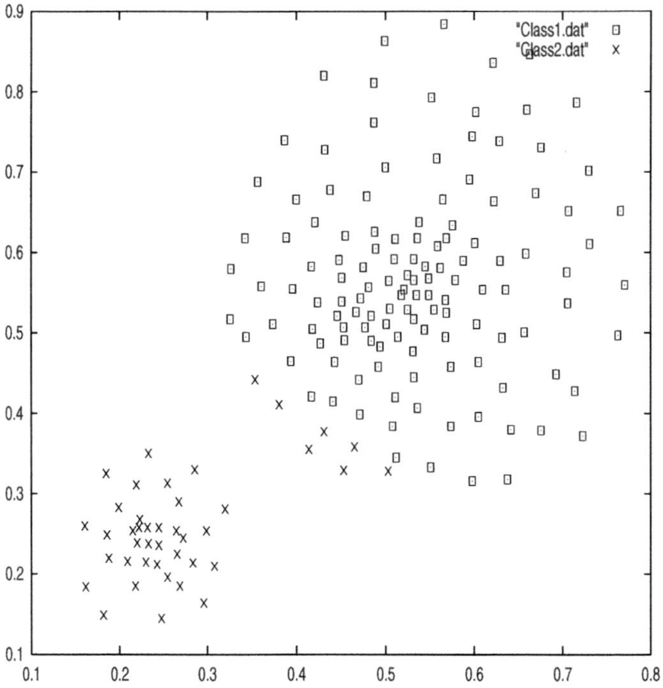

Fig. 3.2. Output from **FCM**

The variable $A = (\alpha_1, \dots, \alpha_c)$ controls cluster sizes. The constraint for A is

$$\mathcal{A} = \{ \; A = (\alpha_1, \dots, \alpha_c) \; : \; \sum_{j=1}^{c} \alpha_j = 1 \; ; \; \alpha_i \geq 0, \; 1 \leq i \leq c \, \}. \tag{3.17}$$

Since we have three variables (U, V, A), the alternate optimization algorithm has a step for A in addition to those for U and V. Notice also that either $J(U, V, A) = J_{\mathrm{efca}}(U, V, A)$ or $J(U, V, A) = J_{\mathrm{fcma}}(U, V, A)$ is used.

Algorithm FCMA: *Fuzzy c-Means with a Variable Controlling Cluster Sizes.*

FCMA1. [Generate initial value:] Generate c initial values for $\bar{V} = (\bar{v}_1, \dots, \bar{v}_c)$ and $\bar{A} = (\bar{\alpha}_1, \dots, \bar{\alpha}_c)$.

FCMA2. [Find optimal U:] Calculate

$$\bar{U} = \arg\min_{U \in \mathcal{U}_f} J(U, \bar{V}, \bar{A}). \tag{3.18}$$

FCMA3. [Find optimal V:] Calculate

$$\bar{V} = \arg\min_{V} J(\bar{U}, V, \bar{A}). \tag{3.19}$$

FCMA4. [Find optimal A:] Calculate

$$\bar{A} = \arg\min_{A \in \mathcal{A}} J(\bar{U}, \bar{V}, A). \tag{3.20}$$

FCMA5. [Test convergence:] If $\bar{U}$ or $\bar{V}$ is convergent, stop; else go to FCMA2.
End FCMA.

We first show solutions for the entropy-based method and then those for the standard method with the additional variable.

3.2.1 Solutions for $J_{\text{efca}}(U, V, A)$

The optimal U, V, and A are respectively given by:

$$u_{ki} = \frac{\alpha_i \exp\left(-\frac{D(x_k, v_i)}{\nu}\right)}{\sum_{j=1}^{c} \alpha_j \exp\left(-\frac{D(x_k, v_j)}{\nu}\right)}, \tag{3.21}$$

$$v_i = \frac{\sum_{k=1}^{N} u_{ki} x_k}{\sum_{k=1}^{N} u_{ki}} \tag{3.22}$$

$$\alpha_i = \frac{\sum_{k=1}^{N} u_{ik}}{n} \tag{3.23}$$

3.2.2 Solutions for $J_{\text{fcma}}(U, V, A)$

The solutions of FCMA2, FCMA3, and FCMA4 are as follows, where $r = \frac{1}{m-1}$.

$$u_{ki} = \left\{ \sum_{j=1}^{c} \left(\frac{\alpha_j}{\alpha_i}\right) \left(\frac{D(x_k, v_i)}{D(x_k, v_j)}\right)^{r} \right\}^{-1} \tag{3.24}$$

$$v_i = \frac{\sum_{k=1}^{N} (u_{ki})^m x_k}{\sum_{k=1}^{N} (u_{ki})^m} \tag{3.25}$$

$$\alpha_i = \left[\sum_{j=1}^{c} \left\{ \frac{\sum_{k=1}^{N} (u_{kj})^m D(x_k, v_j)}{\sum_{k=1}^{N} (u_{ki})^m D(x_k, v_i)} \right\}^{m} \right]^{-1} \tag{3.26}$$

Figure 3.3 shows the result from **FCMA** using J_{efca}; J_{fcma} produces a similar result [111]; we omit the detail.

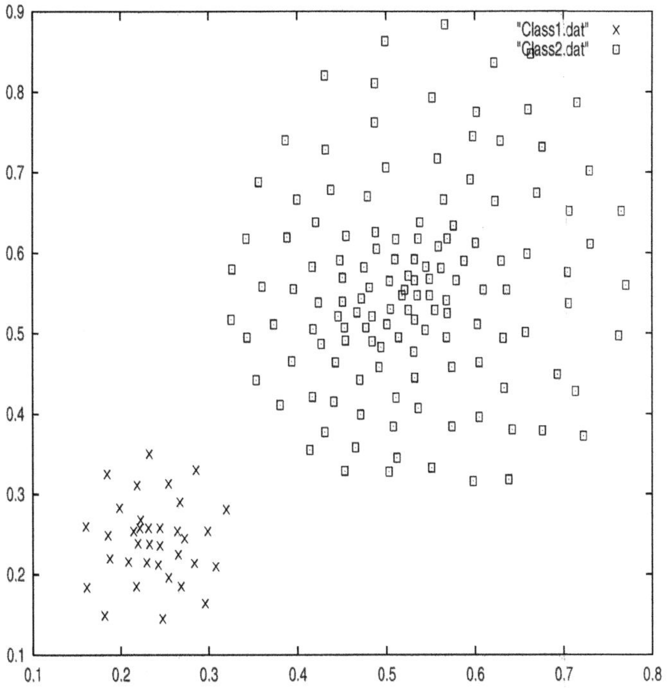

Fig. 3.3. Output from **FCMA** using the entropy method ($\nu^{-1} = 1.2$)

3.3 Covariance Matrices within Clusters

Inclusion of yet another variable is important and indeed has been studied using different algorithms. That is, the use of 'covariance matrices' within clusters. To see the effectiveness of a covariance variable, observe Figure 3.4 where we find two groups, one of which is circular while the other is elongated. A result from **FCM** is shown as Figure 3.5 which fails to separate the two groups. All methods of crisp and fuzzy c-means as well as **FCMA** in the last section fails to separate these groups.

The reason of the failure is that the cluster allocation rule is basically the nearest neighbor allocation, and hence there is no intrinsic rule to recognize the long group to be a cluster.

A remedy is to introduce 'clusterwise Mahalanobis distances'; that is, we consider

$$D(x, v_i; S_i) = (x - v_i)^\top S_i^{-1}(x - v_i), \tag{3.27}$$

where x is assumed to be in cluster i and $S_i = (s_i^{j\ell})$ is a positive definite matrix having p^2 elements $s_i^{j\ell}$, $1 \le j, \ell \le p$. S is used as a variable for another alternate optimization. It will be shown that this variable corresponds to a covariance matrix within cluster i or its fuzzy generalizations. In the following the total set of S_i is denoted by $S = (S_1, S_2, \ldots, S_c)$; S actually has $c \times p^2$ variable elements.

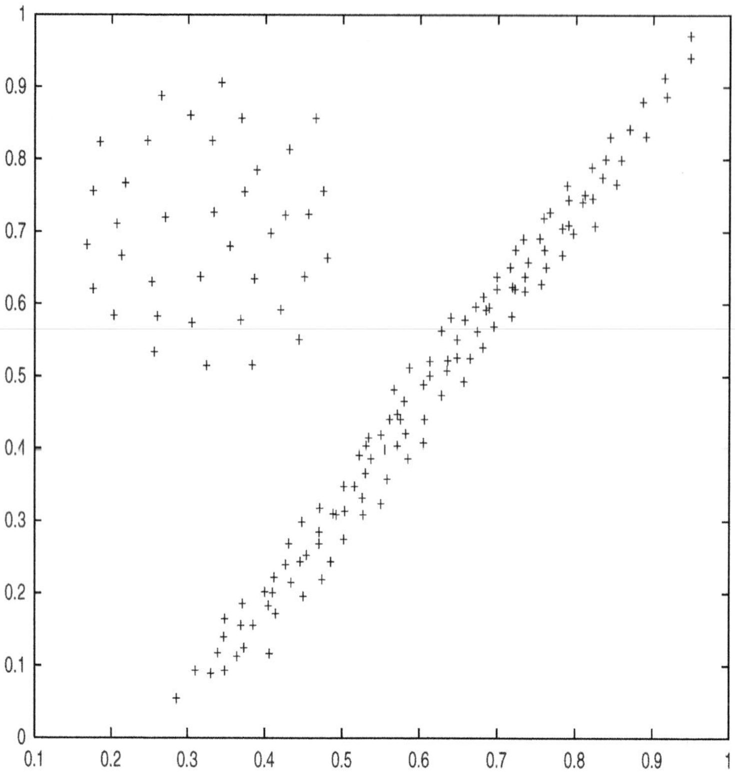

Fig. 3.4. Second artificially generated data set with two groups: one is circular and the other is elongated

We now consider alternate optimization of an objective function with four variables (U, V, A, S). The first method has been proposed by Gustafson and Kessel [45]; the present version includes variable A which is not considered in [45]:

$$J_{\text{fcmas}}(U, V, A, S) = \sum_{i=1}^{c} \sum_{k=1}^{N} (\alpha_i)^{1-m} (u_{ki})^m D(x_k, v_i; S_i) \tag{3.28}$$

with the constraint

$$|S_i| = \rho_i \quad (\rho_i > 0) \tag{3.29}$$

where ρ_i is a fixed parameter and $|S_i|$ is the determinant of S_i.

Accordingly the alternate optimization procedure has the additional step for optimal S.

Algorithm FCMAS: *Fuzzy c-Means with A and S.*

FCMAS1. [Generate initial value:] Generate c initial values for $\bar{V} = (\bar{v}_1, \dots, \bar{v}_c)$, $\bar{A} = (\bar{\alpha}_1, \dots, \bar{\alpha}_c)$, and $\bar{S} = (\bar{S}_1, \bar{S}_2, \dots, \bar{S}_c)$.

FCMAS2. [Find optimal U:] Calculate

$$\bar{U} = \arg\min_{U \in \mathcal{U}_f} J(U, \bar{V}, \bar{A}, \bar{S}). \tag{3.30}$$

FCMAS3. [Find optimal V:] Calculate

$$\bar{V} = \arg\min_{V} J(\bar{U}, V, \bar{A}, \bar{S}). \tag{3.31}$$

FCMAS4. [Find optimal A:] Calculate

$$\bar{A} = \arg\min_{A \in \mathcal{A}} J(\bar{U}, \bar{V}, A, \bar{S}). \tag{3.32}$$

FCMAS5. [Find optimal S:] Calculate

$$\bar{S} = \arg\min_{S} J(\bar{U}, \bar{V}, \bar{A}, S). \tag{3.33}$$

FCMAS6. [Test convergence:] If $\bar{U}$ or $\bar{V}$ is convergent, stop; else go to FCMAS2.
End FCMAS.

Notice that $J = J_{\text{fcmas}}$ in this section.

3.3.1 Solutions for FCMAS by the GK(Gustafson-Kessel) Method

Let us derive the optimal solutions for the GK(Gustafson-Kessel) method. It is evident that the solutions of optimal U, V, and A are the same as those in the last section:

$$u_{ki} = \left\{ \sum_{j=1}^{c} \left(\frac{\alpha_j}{\alpha_i} \right) \left(\frac{D(x_k, v_i; S_i)}{D(x_k, v_j; S_j)} \right)^{r} \right\}^{-1} \tag{3.34}$$

$$v_i = \frac{\sum_{k=1}^{N} (u_{ki})^m x_k}{\sum_{k=1}^{N} (u_{ki})^m} \tag{3.35}$$

$$\alpha_i = \left[\sum_{j=1}^{c} \left\{ \frac{\sum_{k=1}^{N} (u_{kj})^m D(x_k, v_j; S_j)}{\sum_{k=1}^{N} (u_{ki})^m D(x_k, v_i; S_i)} \right\}^{m} \right]^{-1} \tag{3.36}$$

We proceed to derive the optimal S. For this purpose the following Lagrangian function is considered:

$$L = \sum_{i=1}^{c} \sum_{k=1}^{N} (\alpha_i)^{1-m} (u_{ki})^m D(x_k, v_i; S_i) + \sum_{i=1}^{c} \gamma_i \log \frac{|S_i|}{\rho_i}.$$

Differentiation with respect to S_i leads to

$$\frac{\partial L}{\partial S_i} = -\alpha_i^{1-m} \sum_k u_{ki}^m S_i^{-1} (x_k - v_i)(x_k - v_i)^\top S_i^{-1} + \gamma_i S_i^{-1} = 0.$$

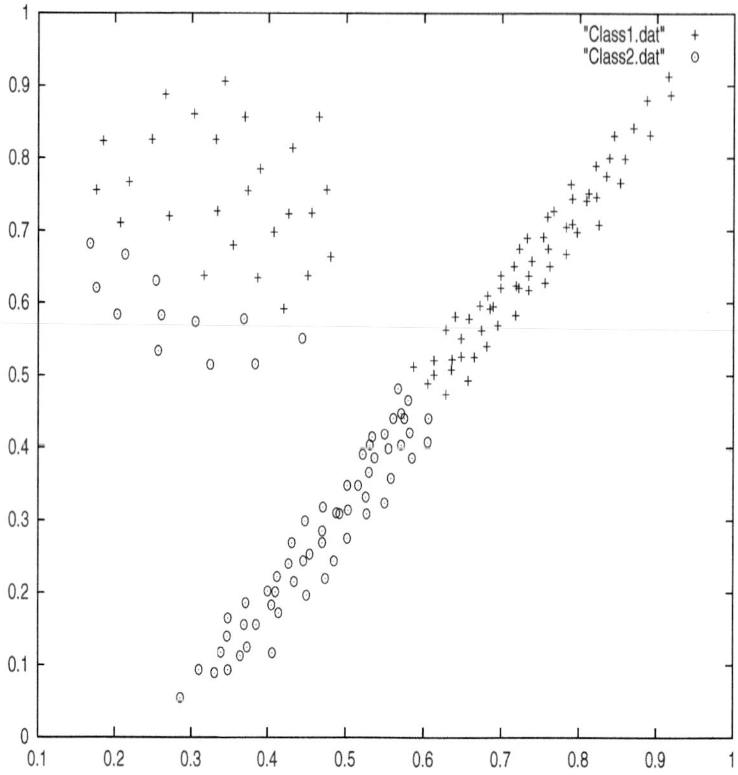

Fig. 3.5. Two clusters by **FCM** for the second artificial data

Putting $\delta_i = \alpha_i^{1-m}/\gamma_i$, we have

$$S_i = \delta_i \sum_{k=1}^{n} u_{ki}^m (x_k - v_i)(x_k - v_i)^\top.$$

To eliminate the Langrange multiplier γ_i, let

$$\hat{S}_i = \sum_{k=1}^{n} u_{ki}^m (x_k - v_i)(x_k - v_i)^\top. \tag{3.37}$$

From the constraint (3.29), we obtain the optimal S_i:

$$S_i = \delta_i \hat{S}_i, \quad \delta_i = \frac{\rho^{\frac{1}{p}}}{|\hat{S}_i|^{\frac{1}{p}}}. \tag{3.38}$$

Note 3.3.1. As to the decision of the parameter ρ_i, Höppner *et al.* [63] suggests $\rho_i = 1$ $(1 \leq i \leq c)$. Other related methods have been studied, e.g., [38]. For more details, readers may refer to [63].

3.4 The KL (Kullback-Leibler) Information Based Method

Ichihashi and his colleagues (e.g., [68]) propose the method of the KL (Kullback-Leibler) information, which uses the next objective function in the algorithm **FCMAS**:

$$J_{\mathrm{KL}}(U,V,A,S)=\sum_{i=1}^{c}\sum_{k=1}^{N}u_{ki}D(x_k,v_i;S_i)+\sum_{i=1}^{c}\sum_{k=1}^{N}u_{ki}\{\nu\log\frac{u_{ik}}{\alpha_i}+\log|S_i|\}. \tag{3.39}$$

3.4.1 Solutions for FCMAS by the Method of KL Information Method

The solutions of optimal U, V, A, and S in the respective step of **FCMAS** with the objective function J_{KL} are as follows.

$$u_{ki}=\frac{\dfrac{\alpha_i}{|S_i|}\exp\left(-\dfrac{D(x_k,v_i;S_i)}{\nu}\right)}{\displaystyle\sum_{j=1}^{c}\frac{\alpha_j}{|S_j|}\exp\left(-\dfrac{D(x_k,v_j;S_i)}{\nu}\right)}, \tag{3.40}$$

$$v_i=\frac{\displaystyle\sum_{k=1}^{N}u_{ki}x_k}{\displaystyle\sum_{k=1}^{N}u_{ki}} \tag{3.41}$$

$$\alpha_i=\frac{\displaystyle\sum_{k=1}^{N}u_{ik}}{n} \tag{3.42}$$

$$S_i=\frac{1}{\displaystyle\sum_{k=1}^{N}u_{ki}}\sum_{k=1}^{N}u_{ki}(x_k-v_i)(x_k-v_i)^{\top} \tag{3.43}$$

Figure 3.6 shows a result of two clusters by the KL method. Such a successful separation of the groups can be attained by the GK method as well.

Note 3.4.1. Let us derive the solution of S_i. From

$$\frac{\partial J_{\mathrm{KL}}}{\partial S_i}=-\sum_{j=1}^{N}u_{ki}S_i^{-1}(x_k-v_i)(x_k-v_i)^{\top}S_i^{-1}+\sum_{j=1}^{N}u_{ki}S_i^{-1}=0,$$

we have (3.43).

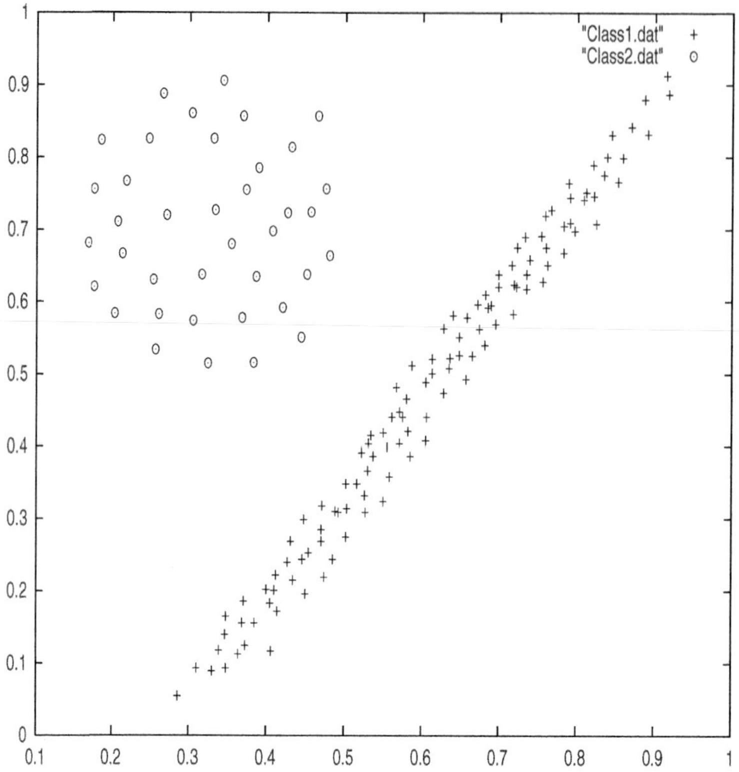

Fig. 3.6. Two clusters by **FCMAS** by the KL method for the second artificial data

Note 3.4.2. Readers can observe the solution by the KL method is similar to that by the EM algorithm for the mixture of normal distributions given in the previous chapter. Although the matrix S_i in the KL method or GK method is different from the 'true covariance' within a cluster in the mixture of normal distributions, we can compare S_i in the KL and GK methods to a representation of covariance within a cluster by analogy.

3.5 Defuzzified Methods of c-Means Clustering

We have seen how methods of fuzzy c-means have been derived by introducing nonlinear terms into the objective function for crisp c-means. In this section we consider the converse, that is, linearization of objective functions of variations of fuzzy c-means, whereby we attempt to derive new algorithms of variations of the crisp c-means. This process to derive an algorithm of crisp c-means is called defuzzification of fuzzy c-means after the corresponding name in fuzzy control. Concretely, we intend to introduce the size variable and the 'covariance' variable into methods of crisp c-means.

For this purpose the algorithm **FCMA** with J_{efca} and **FCMAS** with J_{KL} are useful. Recall the two objective functions:

$$J_{\mathrm{efca}}(U,V,A)=\sum_{i=1}^{c}\sum_{k=1}^{N}u_{ki}D(x_k,v_i)+\nu\sum_{i=1}^{c}\sum_{k=1}^{N}\{u_{ki}\log u_{ki}-u_{ki}\log\alpha_i\},$$

$$J_{\mathrm{KL}}(U,V,A,S)=\sum_{i=1}^{c}\sum_{k=1}^{N}u_{ki}D(x_k,v_i;S_i)$$
$$+\sum_{i=1}^{c}\sum_{k=1}^{N}\{\nu u_{ki}\log u_{ik}-\nu u_{ki}\log\alpha_i+u_{ki}\log|S_i|\}.$$

If we eliminate the term of entropy $u_{ki}\log u_{ki}$ from the both functions, the objective functions are linearized with respect to u_{ki}:

$$J_{\mathrm{defc}}(U,V,A)=\sum_{i=1}^{c}\sum_{k=1}^{N}u_{ki}D(x_k,v_i)-\nu\sum_{i=1}^{c}\sum_{k=1}^{N}u_{ki}\log\alpha_i, \tag{3.44}$$

$$J_{\mathrm{dKL}}(U,V,A,S)=\sum_{i=1}^{c}\sum_{k=1}^{N}u_{ki}D(x_k,v_i;S_i) \tag{3.45}$$
$$+\sum_{i=1}^{c}\sum_{k=1}^{N}\{-\nu u_{ki}\log\alpha_i+u_{ki}\log|S_i|\}. \tag{3.46}$$

We use (3.44) and (3.46) in **FCMA** and **FCMAS**, respectively.

3.5.1 Defuzzified c-Means with Cluster Size Variable

Let us employ J_{defc} in **FCMA**. We then obtain the following optimal solutions.

$$u_{ki}=\begin{cases}1 & (i=\arg\min_{1\le j\le c}D_{jk}-\nu\log\alpha_j),\\ 0 & (\text{otherwise}),\end{cases}$$

$$v_i=\frac{\sum_{k=1}^{N}u_{ki}x_k}{\sum_{k=1}^{N}u_{ki}},$$

$$\alpha_i=\frac{1}{N}\sum_{k=1}^{N}u_{ki}.$$

If we define

$$G_i=\{x_k : u_{ki}=1\},$$

we have

$$v_i = \frac{\sum_{x_k \in G_i} x_k}{|G_i|}, \tag{3.47}$$

$$\alpha_i = \frac{|G_i|}{N} \tag{3.48}$$

where $|G_i|$ is the number of elements in G_i.

3.5.2 Defuzzification of the KL-Information Based Method

The optimal solutions when we use J_{dKL} in **FCMAS** are as follows.

$$u_{ki} = \begin{cases} 1 & (i = \arg\min_{1 \le j \le c} D_{jk} - \nu \log \alpha_j + \log |S_j|), \\ 0 & (\text{otherwise}). \end{cases}$$

$$v_i = \frac{\sum_{k=1}^{N} u_{ik} x_k}{\sum_{k=1}^{N} u_{ik}},$$

$$\alpha_i = \frac{1}{N} \sum_{k=1}^{n} u_{ik},$$

$$S_i = \frac{1}{\sum_{k=1}^{N} u_{ik}} \sum_{k=1}^{N} u_{ik} (x_k - v_i)(x_k - v_i)^\top,$$

We note the solutions for V and A are the same as (3.47) and (3.48), respectively.

3.5.3 Sequential Algorithm

In section 2.2 we have considered a sequential algorithm of crisp c-means as algorithm **CMS**. We consider the variation of **CMS** based on the defuzzified objective function J_{dKL}. The algorithm is as follows.

Algorithm KLCMS.

KLCMS1. Set initial clusters.
KLCMS2. For all x_k, repeat KLCMS2-1 and KLCMS2-2.
  KLCMS2-1. Allocate x_k to the cluster that minimizes

$$D_{jk} - \nu \log \alpha_j + \log |S_j|.$$

  KLCMS2-2. If membership of x_k is changed from cluster (say q) to another cluster (say r), update variables V, α, and S.
KLCMS3. If the solution is convergent, stop; else go to KLCMS2.
End of KLCMS.

3.5.4 Efficient Calculation of Variables

Assume that a variable like v' with prime denotes that after the update, while that without prime, like v, is before the update. Suppose that x_k moves from cluster q to cluster r. N_q and N_r are the numbers of elements in clusters q and r, respectively.

First, the center is easily updated:

$$v'_q = \frac{N_q}{N_q - 1} v_q - \frac{x_k}{N_q - 1},$$

$$v'_r = \frac{N_r}{N_r + 1} v_r + \frac{x_k}{N_r + 1}$$

To update α is also easy:

$$\alpha'_q = \alpha_q - \frac{1}{N} \quad , \quad \alpha'_r = \alpha_r + \frac{1}{N}$$

To update covariance matrices efficiently requires a more sophisticated technique. For this purpose we use the Sherman-Morrison-Woodbury formula [43].

Put

$$B_r = \sum_{G_r} x x^\top,$$

and note

$$S_r = \frac{B_r}{N_r} - v_r v_r^\top. \tag{3.49}$$

Let S'_r be the covariance matrix after this move. Then,

$$S'_r = \frac{B'_r}{N_r + 1} - v'_r v'^T_r \tag{3.50}$$

Notice also

$$\begin{aligned} S_r^{-1} &= (\frac{B_r}{N_r} - v_r v_r^\top)^{-1} \\ &= N_r B_r^{-1} + \frac{N_r^2}{1 - N_r v_r^\top B_r^{-1} v_r} B_r^{-1} v_r v_r^\top B_r^{-1}. \end{aligned}$$

Using

$$\begin{aligned} B_r'^{-1} &= (B_r + x_k x_k^\top)^{-1} \\ &= B_r^{-1} - \frac{1}{x_k^\top B_r^{-1} x_k + 1} B_r^{-1} x_k x_k^\top B_r^{-1}, \end{aligned}$$

we can update $S_r'^{-1}$.

Note 3.5.1. When **FCMAS** using J_{dKL} is applied to Figure 3.4, the result is the same as that in Figure 3.6, i.e., the two groups are successfully separated, whereas **FCMA** using J_{defc} cannot separate these groups. For the first example in Figure 3.1, the both methods separate the two groups without misclassifications as in Figure 3.2. Notice also that the latter method without S is less time-consuming than the former method with the calculation of S.

3.6 Fuzzy c-Varieties

There are other variations of fuzzy c-means of which most frequently applied are fuzzy c-varieties (FCV) [6] and fuzzy c-regression models (FCRM) [48]. We describe them in this and next sections.

We generalize the dissimilarity $D(x_k, v_i)$ for this purpose. Instead of the dissimilarity between object x_k and cluster center v_i, we consider $D(x_k, P_i)$, dissimilarity between x_k and a prototype of cluster i, where the meaning of P_i varies according to the method of clustering. Thus, $P_i = v_i$ in fuzzy c-means clustering, but P_i describes a lower-dimensional hyperplane, and $D(x_k, P_i)^{\frac{1}{2}}$ is the distance between x_k and the hyperplane in FCV; P_i describes a regression surface in FCRM. To simplify the notation, we sometimes write D_{ki} instead of $D(x_k, P_i)$ when the simplified symbol is more convenient for our purpose.

Let us consider fuzzy c-varieties. For simplicity we first consider case when the linear variety is one-dimensional, and then extend to the case of multidimensional hyperplane.

We consider the next objective functions:

$$J_{\text{fcv}}(U, P) = \sum_{i=1}^{c}\sum_{k=1}^{N}(u_{ki})^m D(x_k, P_i) = \sum_{i=1}^{c}\sum_{k=1}^{N}(u_{ki})^m D_{ki}, \tag{3.51}$$

$$\begin{aligned} J_{\text{efcv}}(U, P) &= \sum_{i=1}^{c}\sum_{k=1}^{N}\{u_{ki} D(x_k, P_i) + \nu u_{ki}\log u_{ki}\} \\ &= \sum_{i=1}^{c}\sum_{k=1}^{N}\{u_{ki} D_{ki} + \nu u_{ki}\log u_{ki}\} \end{aligned} \tag{3.52}$$

where $P = (P_1, \dots, P_c)$ is the collection of all prototypes for clusters $1, \dots, c$. We use either $J = J_{\text{fcv}}$ or $J = J_{\text{efcv}}$ in the alternate optimization algorithm of FCV which will be shown below.

In the one-dimensional case, the linear variety is the line described by two vectors $P_i = (w_i, t_i)$ in which t_i represents the direction of the line and $\|t_i\| = 1$, whereby the line is described by

$$\ell(\beta) = w_i + \beta t_i, \quad \beta \in \boldsymbol{R}.$$

The squared distance between x_k and $\ell(\beta)$ is

$$\min_{\beta \in \boldsymbol{R}} \|x_k - \ell(\beta)\|^2 = \|x_k - w_i\|^2 - \langle x_k - w_i, t_i\rangle^2$$

where $\langle x, y\rangle$ is the inner product of x and y. We therefore define

$$D_{ki} = D(x_k, P_i) = \|x_k - w_i\|^2 - \langle x_k - w_i, t_i\rangle^2 \tag{3.53}$$

for the one-dimensional case.

As expected, the next algorithm **FCV** itself is a simple rewriting of the **FCM** algorithm.

Algorithm FCV: *Fuzzy c-Varieties.*

FCV1. [Generate initial value:] Generate c initial prototypes $\bar{P} = (\bar{P}_1, \dots, \bar{P}_c)$.
FCV2. [Find optimal U:] Calculate

$$\bar{U} = \arg\min_{U \in \mathcal{U}_f} J(U, \bar{P}). \tag{3.54}$$

FCV3. [Find optimal P:] Calculate

$$\bar{P} = \arg\min_{P} J(\bar{U}, P). \tag{3.55}$$

FCV4. [Test convergence:] If $\bar{U}$ is convergent, stop; else go to FCV2.
End FCV.

in which either $J = J_{\text{fcv}}$ or $J = J_{\text{efcv}}$.

It is obvious that the optimal solution of U is obtained by just rewriting those for fuzzy c-means, that is,

$$\bar{u}_{ki} = \left[\sum_{j=1}^{c} \left(\frac{D(x_k, \bar{P}_i)}{D(x_k, \bar{P}_j)}\right)^{\frac{1}{m-1}}\right]^{-1} = \left[\sum_{j=1}^{c} \left(\frac{D_{ki}}{D_{kj}}\right)^{\frac{1}{m-1}}\right]^{-1}, \tag{3.56}$$

$$\bar{u}_{ki} = \frac{\exp\left(-\dfrac{D(x_k, \bar{P}_i)}{\nu}\right)}{\displaystyle\sum_{j=1}^{c} \exp\left(-\dfrac{D(x_k, \bar{P}_j)}{\nu}\right)} = \frac{\exp\left(-\dfrac{D_{ki}}{\nu}\right)}{\displaystyle\sum_{j=1}^{c} \exp\left(-\dfrac{D_{kj}}{\nu}\right)}, \tag{3.57}$$

for $J = J_{\text{fcv}}$ and $J = J_{\text{efcv}}$, respectively.

It is known [6] that the optimal solution of P using $J = J_{\text{fcv}}$ are given by the following.

$$w_i = \frac{\displaystyle\sum_{k=1}^{N} (u_{ki})^m x_k}{\displaystyle\sum_{k=1}^{N} (u_{ki})^m}, \tag{3.58}$$

while t_i is the normalized eigenvector corresponding to the maximum eigenvalue of the matrix

$$A_i = \sum_{k=1}^{N} (u_{ki})^m (x_k - w_i)(x_k - w_i)^\top. \tag{3.59}$$

When $J = J_{\text{efcv}}$ is used, we have

$$w_i = \frac{\displaystyle\sum_{k=1}^{N} u_{ki} x_k}{\displaystyle\sum_{k=1}^{N} u_{ki}}, \tag{3.60}$$

and t_i is the normalized eigenvector corresponding to the maximum eigenvalue of the matrix

$$A_i = \sum_{k=1}^{N} u_{ki}(x_k - w_i)(x_k - w_i)^\top. \tag{3.61}$$

3.6.1 Multidimensional Linear Varieties

In the multidimensional case, let the dimension of the linear variety be $q > 1$. Assume the variety $\boldsymbol{L}_i^q$ for cluster i is represented by the center w_i and normalized vectors $t_{i1}, \ldots, t_{iq}$. Then every $y \in \boldsymbol{L}_i^q$ is represented as

$$y = w_i + \sum_{\ell=1}^{q} \beta_\ell t_{i\ell}, \quad \beta_1, \ldots, \beta_q \in \boldsymbol{R}.$$

Hence the dissimilarity is

$$D_{ki} = D(x_k, P_i) = \|x_k - w_i\|^2 - \sum_{\ell=1}^{q} \langle x_k - w_i, t_{i\ell} \rangle^2. \tag{3.62}$$

Then the solution for FCV2 is given by the same equation (3.56) (or (3.57) in the case of entropy-based method), and that for FCV3 is (3.58) (or (3.60)).

It is also known [6] that the optimal $t_{i1}, \ldots, t_{iq}$ are given by the q eigenvectors corresponding to q largest eigenvalues of the same matrix

$$A_i = \sum_{k=1}^{N} (u_{ki})^m (x_k - w_i)(x_k - w_i)^\top.$$

If the entropy-based method is used, $m = 1$ in the above matrix A_i. It should also be noted that the q eigenvectors should be orthogonalized and normalized.

3.7 Fuzzy c-Regression Models

To obtain clusters and the corresponding regression models have been proposed by Hathaway and Bezdek [48]. In order to describe this method, we assume a data set $\{(x_1, y_1), \ldots, (x_N, y_N)\}$ in which $x_1, \ldots, x_N \in \boldsymbol{R}^p$ are data of the independent variable x and $y_1, \ldots, y_N \in \boldsymbol{R}$ are those of the dependent variable y.

What we need to have is the c regression models:

$$y = f_i(x; \beta_i) + e_i, \quad i = 1, \ldots, c.$$

We assume the regression models to be linear:

$$f_i(x; \beta_i) = \sum_{j=1}^{p} \beta_i^j x^j + \beta_i^{p+1}.$$

for simplicity. We moreover put

$$z = (x, 1)^\top = (x^1, \dots, x^p, 1)^\top$$

and accordingly

$$z_k = (x_k, 1)^\top = (x_k^1, \dots, x_k^p, 1)^\top,$$
$$\beta_i = (\beta_i^1, \dots, \beta_i^{p+1})$$

in order to simplify the derivation.

Since the objective of a regression model is to minimize the error e_i^2, we define

$$D_{ki} = D((x_k, y_k), \beta_i)) = (y_k - \sum_{j=1}^{p} \beta_i^j x^j - \beta_i^{p+1})^2, \tag{3.63}$$

or in other symbols,

$$D_{ki} = (y_k - \langle z_k, \beta_i \rangle)^2.$$

We consider the next objective functions:

$$J_{\text{fcrm}}(U, B) = \sum_{i=1}^{c} \sum_{k=1}^{N} (u_{ki})^m D((x_k, y_k), \beta_i) = \sum_{i=1}^{c} \sum_{k=1}^{N} (u_{ki})^m D_{ki}, \tag{3.64}$$

$$J_{\text{efcrm}}(U, B) = \sum_{i=1}^{c} \sum_{k=1}^{N} \{u_{ki} D((x_k, y_k), \beta_i) + \nu u_{ki} \log u_{ki}\}$$
$$= \sum_{i=1}^{c} \sum_{k=1}^{N} \{u_{ki} D_{ki} + \nu u_{ki} \log u_{ki}\} \tag{3.65}$$

where $B = (\beta_1, \dots, \beta_c)$.

Algorithm FCRM: *Fuzzy c-Regression Models.*

FCRM1. [Generate initial value:] Generate c initial prototypes $\bar{B} = (\bar{\beta}_1, \dots, \bar{\beta}_c)$.
FCRM2. [Find optimal U:] Calculate

$$\bar{U} = \arg \min_{U \in \mathcal{U}_f} J(U, \bar{B}). \tag{3.66}$$

FCRM3. [Find optimal B:] Calculate

$$\bar{B} = \arg \min_{B} J(\bar{U}, B). \tag{3.67}$$

FCRM4. [Test convergence:] If $\bar{U}$ is convergent, stop; else go to FCRM2.
End FCRM.

in which either $J = J_{\text{fcrm}}$ or $J = J_{\text{efcrm}}$.

The optimal solution for U is

$$\bar{u}_{ki} = \left[\sum_{j=1}^{c} \left(\frac{D_{ki}}{D_{kj}} \right)^{\frac{1}{m-1}} \right]^{-1}, \tag{3.68}$$

$$\bar{u}_{ki} = \frac{\exp\left(-\frac{D_{ki}}{\nu}\right)}{\sum_{j=1}^{c} \exp\left(-\frac{D_{kj}}{\nu}\right)}, \tag{3.69}$$

for $J = J_{\mathrm{fcrm}}$ and $J = J_{\mathrm{efcrm}}$, respectively.

The derivation of the optimal solution for B is not difficult. From $\frac{\partial J}{\partial \beta_i^j} = 0$, we have

$$\left(\sum_{k=1}^{n} (u_{ik})^m z_k z_k^\top \right) \beta_i = \sum_{k=1}^{n} (u_{ik})^m y_k z_k$$

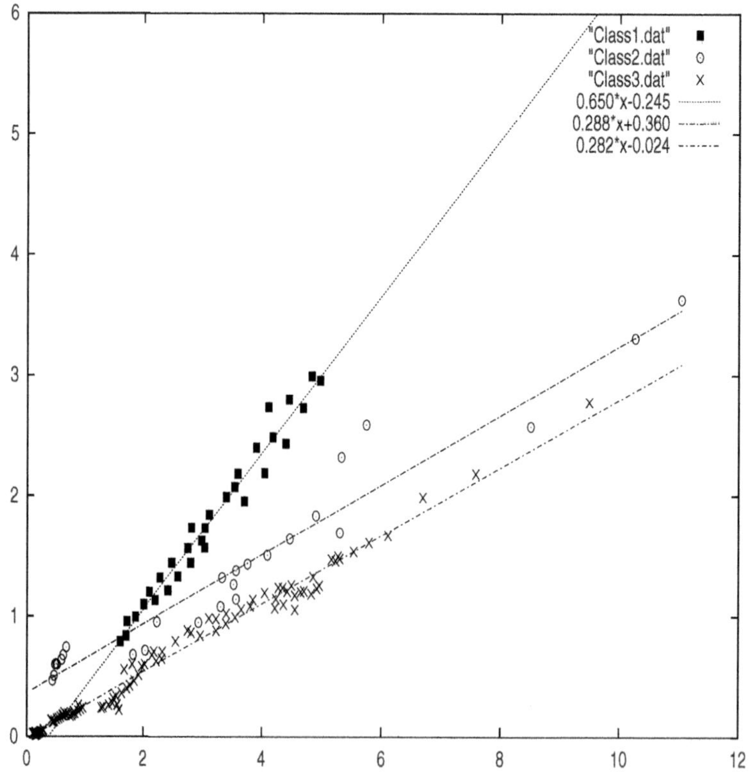

Fig. 3.7. Clusters when **FCRM** was applied to the sample data set ($c = 3$)

Hence the solution is

$$\beta_i = \left(\sum_{k=1}^{n} (u_{ik})^m z_k z_k^\top \right)^{-1} \sum_{k=1}^{n} (u_{ik})^m y_k z_k. \tag{3.70}$$

In the case of the entropy-based method, we put $m = 1$ in (3.70).

Note 3.7.1. For the regression models, the additional variables A and S can be included. Since the generalization is not difficult, we omit the detail.

Figure 3.7 shows a result from **FCRM** using J_{efcrm}. The number of clusters is three. The data set is based on actual data on energy consumption of different countries in Asia in different years, but the data source is not made public. We observe that the result is rather acceptable from the viewpoint of clustering and regression models, but a cluster in the middle is not well-separated from others.

3.8 Noise Clustering

Davé [22, 23] proposes the method of noise clustering. His idea is simple and useful in many applications. Hence we overview the method of noise clustering in this section. Since his method is based on the objective function by Dunn and Bezdek, we assume

$$J_{\mathrm{fcm}}(U, V) = \sum_{i=1}^{c} \sum_{k=1}^{N} (u_{ki})^m D(x_k, v_i).$$

Let us add another cluster $c+1$ in which there is no center and the dissimilarity $D_{k,c+1}$ between x_k and this cluster be

$$D_{k,c+1} = \delta$$

where $\delta > 0$ is a fixed parameter. We thus consider the objective function

$$J_{\mathrm{nfcm}}(U, V) = \sum_{i=1}^{c} \sum_{k=1}^{N} (u_{ki})^m D(x_k, v_i) + \sum_{k=1}^{N} (u_{k,c+1})^m \delta. \tag{3.71}$$

where U is $N \times (c+1)$ matrix, while $V = (v_1, \ldots, v_c)$. Notice also that the constraint is

$$\mathcal{U}_f = \{ U = (u_{ki}) : \sum_{j=1}^{c+1} u_{kj} = 1,\ 1 \le k \le N;$$
$$u_{ki} \in [0, 1],\ 1 \le k \le N,\ 1 \le i \le c+1 \}.$$

When this objective function is used in algorithm **FCM**, the optimal solutions are given by the following.

$$\bar{u}_{ki} = \left[\sum_{j=1}^{c} \left(\frac{D(x_k, \bar{v}_i)}{D(x_k, \bar{v}_j)} \right)^{\frac{1}{m-1}} + \left(\frac{D(x_k, \bar{v}_i)}{\delta} \right)^{\frac{1}{m-1}} \right]^{-1},\ 1 \le i \le c \tag{3.72}$$

$$\bar{u}_{k,c+1} = \left[\sum_{j=1}^{c} \left(\frac{\delta}{D(x_k, \bar{v}_j)} \right)^{\frac{1}{m-1}} + 1 \right]^{-1}, \tag{3.73}$$

$$\bar{v}_i = \frac{\sum_{k=1}^{N} (\bar{u}_{ki})^m x_k}{\sum_{k=1}^{N} (\bar{u}_{ki})^m}, \quad 1 \le i \le c. \tag{3.74}$$

If we use the entropy-based objective function

$$J_{\text{nefcm}}(U, V) = \sum_{i=1}^{c} \sum_{k=1}^{N} u_{ki} D(x_k, v_i) + \sum_{k=1}^{N} u_{k,c+1} \delta + \nu \sum_{i=1}^{c+1} \sum_{k=1}^{N} u_{ki} \log u_{ki} \tag{3.75}$$

in algorithm **FCM**, we have the next solutions.

$$\bar{u}_{ki} = \frac{\exp\left(-\frac{D(x_k, \bar{v}_i)}{\nu} \right)}{\sum_{j=1}^{c} \exp\left(-\frac{D(x_k, \bar{v}_j)}{\nu} \right) + \exp\left(-\frac{\delta}{\nu} \right)}, \quad 1 \le i \le c, \tag{3.76}$$

$$\bar{u}_{k,c+1} = \frac{\exp\left(-\frac{\delta}{\nu} \right)}{\sum_{j=1}^{c} \exp\left(-\frac{D(x_k, \bar{v}_j)}{\nu} \right) + \exp\left(-\frac{\delta}{\nu} \right)}, \tag{3.77}$$

$$\bar{v}_i = \frac{\sum_{k=1}^{N} \bar{u}_{ki} x_k}{\sum_{k=1}^{N} \bar{u}_{ki}}, \quad 1 \le i \le c. \tag{3.78}$$

4 Variations and Generalizations - II

This chapter continues to describe various generalizations and variations of fuzzy c-means clustering. The methods studied here are more specific or include more recent techniques. In a sense some of them are more difficult to understand than those in the previous section. It does not imply, however, that methods described in this chapter are less useful.

4.1 Kernelized Fuzzy c-Means Clustering and Related Methods

Recently support vector machines [163, 164, 14, 20] have been remarked by many researchers. Not only support vector machines themselves but also related techniques employed in them have also been noted and found to be useful. A typical example of these techniques is the use of kernel functions [163, 14] which enables nonlinear classifications, i.e., a classifier which has nonlinear boundaries for different classes.

In supervised classification problems, to generate nonlinear boundaries itself is simple: just apply the nearest neighbor classification [29, 30], or by techniques in neural networks such as the radial basis functions [10]. In contrast, techniques to generate nonlinear boundaries in clustering are more limited. It is true that we have boundaries of quadratic functions by the GK and KL methods in the previous chapter, but we cannot obtain highly nonlinear boundaries as those by kernel functions [163, 14] in support vector machines.

Two methods related to support vector machines and clusters with nonlinear boundaries have been studied. Ben-Hur *et al.* [3, 4] proposed a support vector clustering method and they showed clusters with highly nonlinear boundaries. Their method employs a variation of an algorithm in support vector machines; a quadratic programming technique and also additional calculation to find clusters are necessary. Although they show highly nonlinear cluster boundaries, their algorithm is complicated and seems time-consuming. Another method has been proposed by Girolami [42] who simply uses a kernel function in c-means with a stochastic approximation algorithm without reference to support vector machines.

S. Miyamoto et al.: Algorithms for Fuzzy Clustering, STUDFUZZ 229, pp. 67–98, 2008.
springerlink.com

The latter type of algorithms is useful when we discuss fuzzy c-means with kernel functions [108], which we study in this section.

4.1.1 Transformation into High-Dimensional Feature Space

Let us briefly review how a kernel function is used for classification in general. Suppose a set of objects $x_1, \ldots, x_N \in \boldsymbol{R}^p$ should be analyzed, but structure in the original space $\boldsymbol{R}^p$ is somehow inadequate for the analysis. The idea of the method of kernel functions (also called kernel trick) uses a 'nonlinear' transformation $\Phi\colon \boldsymbol{R}^p \to H$, where H is a high-dimensional Euclidean space, which is sometimes an infinite dimensional Hilbert space [94]. We can write

$$\Phi(x) = (\phi_1(x), \phi_2(x), \ldots).$$

Notice that the number of components as well as components $\phi_\ell(x)$ themselves in the above expression are not important, as we see below. Moreover, it is unnecessary to have a functional form of $\Phi(x)$. Instead, we assume that the scalar product of H is known and is given by a known kernel function $K(x, y)$:

$$K(x, y) = \langle \Phi(x), \Phi(y) \rangle, \tag{4.1}$$

where $\langle \cdot, \cdot \rangle$ in (4.1) is the scalar product of H. It is sometimes written as $\langle \cdot, \cdot \rangle_H$ with the explicit index of H. We emphasize again that by the kernel trick we do not know $\Phi(x)$ but we know $K(x, y) = \langle \Phi(x), \Phi(y) \rangle$.

Using the idea of kernel functions, we are capable of analyzing objects in space H instead of the original space $\boldsymbol{R}^p$; $\boldsymbol{R}^p$ is called data space while H is called a (high-dimensional) feature space. Several types of analysis including regression, principal component, classification, and clustering have been done using the feature space with kernels (see e.g., [145]). In applications, two types of kernel functions are most frequently used:

$$K(x, y) = (\langle x, y \rangle + c)^d, \tag{4.2}$$

$$K(x, y) = \exp\left(-\lambda \|x - y\|^2\right). \tag{4.3}$$

The former is called a polynomial kernel and the latter is called a Gaussian kernel. We use the Gaussian kernel below.

Kernelized fuzzy c-means algorithm. Recall that the objects for clustering are $x_k = (x_k^1, \ldots, x_k^p) \in \boldsymbol{R}^p$, $(k = 1, 2, \ldots, N)$ where $\boldsymbol{R}^p$ is the p-dimensional Euclidean space with the scalar product $\langle x, y \rangle = x^\top y$. The method of fuzzy c-means uses the alternate optimization of

$$J_{\text{fcm}}(U, V) = \sum_{i=1}^{c} \sum_{k=1}^{N} (u_{ki})^m \|x_k - v_i\|^2$$

or the entropy-based function

$$J_{\text{efc}}(U, V) = \sum_{i=1}^{c} \sum_{k=1}^{N} \{u_{ki} \|x_k - v_i\|^2 + \nu u_{ki} \log u_{ki}\}.$$

If we intend to analyze data using a kernel function, we should transform data into the high-dimensional feature space, in other words, transformed objects $\Phi(x_1), \ldots, \Phi(x_N)$ should be divided into clusters.

The following objective functions should therefore be considered.

$$J_{\mathrm{kfcm}}(U, W) = \sum_{i=1}^{c} \sum_{k=1}^{N} (u_{ki})^m \|\Phi(x_k) - W_i\|_H^2, \tag{4.4}$$

$$J_{\mathrm{kefc}}(U, W) = \sum_{i=1}^{c} \sum_{k=1}^{N} \{u_{ki} \|\Phi(x_k) - W_i\|_H^2 + \nu u_{ki} \log u_{ki}\}, \tag{4.5}$$

where $W = (W_1, \ldots, W_c)$ and W_i is the cluster center in H; $\|\cdot\|_H$ is the norm of H [94]. Notice that H is abstract and its elements do not have explicit representation; indeed, there are examples in which H is not uniquely determined while the scalar product is uniquely specified [14]. It should hence be noticed that we cannot have $\Phi(x_k)$ and W_i explicitly.

The alternate optimization algorithm **FCM** should be applied to (4.4) and (4.5). The optimal U is

$$\begin{aligned} \bar{u}_{ki} &= \left[\sum_{j=1}^{c} \left(\frac{\|\Phi(x_k) - \bar{W}_i\|_H^2}{\|\Phi(x_k) - \bar{W}_j\|_H^2} \right)^{\frac{1}{m-1}} \right]^{-1} \\ &= \left[\sum_{j=1}^{c} \left(\frac{D_{ki}}{D_{kj}} \right)^{\frac{1}{m-1}} \right]^{-1}, \end{aligned} \tag{4.6}$$

$$\begin{aligned} \bar{u}_{ki} &= \frac{\exp\left(-\dfrac{\|\Phi(x_k) - \bar{W}_i\|_H^2}{\nu} \right)}{\displaystyle\sum_{j=1}^{c} \exp\left(-\dfrac{\|\Phi(x_k) - \bar{W}_j\|_H^2}{\nu} \right)} \\ &= \frac{\exp\left(-\dfrac{D_{ki}}{\nu} \right)}{\displaystyle\sum_{j=1}^{c} \exp\left(-\dfrac{D_{kj}}{\nu} \right)} \end{aligned} \tag{4.7}$$

for $J_{\mathrm{kfcm}}(U, W)$ and $J_{\mathrm{kefc}}(U, W)$, respectively. Notice that we put $D_{ki} = \|\Phi(x_k) - \bar{W}_i\|_H^2$.

The optimal W is given by

$$\bar{W}_i = \frac{\displaystyle\sum_{k=1}^{N} (\bar{u}_{ki})^m \Phi(x_k)}{\displaystyle\sum_{k=1}^{N} (\bar{u}_{ki})^m}. \tag{4.8}$$

$$\bar{W}_i = \frac{\sum_{k=1}^{N} \bar{u}_{ki} \Phi(x_k)}{\sum_{k=1}^{N} \bar{u}_{ki}}. \tag{4.9}$$

for $J_{\text{kfcm}}(U, W)$ and $J_{\text{kefc}}(U, W)$, respectively. Notice that these equations should be derived from calculus of variations in the abstract Hilbert space H. For the derivation, see the note below.

Notice that these formulas cannot directly be used since the explicit form of $\Phi(x_k)$ (and hence $\bar{W}_i$) is unknown. To solve this problem, we substitute the solution (4.8) into $D_{ki} = \|\Phi(x_k) - \bar{W}_i\|_H^2$. Then the next equation holds.

$$\begin{aligned} D_{ki} &= \langle \Phi(x_k) - \bar{W}_i, \Phi(x_k) - \bar{W}_i \rangle \\ &= \langle \Phi(x_k), \Phi(x_k) \rangle - 2\langle \bar{W}_i, \Phi(x_k) \rangle + \langle \bar{W}_i, \bar{W}_i \rangle \\ &= \langle \Phi(x_k), \Phi(x_k) \rangle - \frac{2}{S_i(m)} \sum_{j=1}^{N} (u_{ji})^m \langle \Phi(x_j), \Phi(x_k) \rangle \\ &+ \frac{1}{S_i(m)^2} \sum_{j=1}^{N} \sum_{\ell=1}^{N} u_{ji} u_{\ell i} \langle \Phi(x_j), \Phi(x_\ell) \rangle, \end{aligned}$$

where

$$S_i(m) = \sum_{k=1}^{N} (\bar{u}_{ki})^m. \tag{4.10}$$

Let

$$K_{k\ell} = K(x_k, x_\ell) = \langle \Phi(x_k), \Phi(x_\ell) \rangle.$$

We then have

$$D_{ki} = K_{kk} - \frac{2}{S_i(m)} \sum_{j=1}^{N} (u_{ji})^m K_{jk} + \frac{1}{S_i(m)^2} \sum_{j=1}^{N} \sum_{\ell=1}^{N} (u_{ji} u_{\ell i})^m K_{j\ell}. \tag{4.11}$$

Notice also that when the entropy method is used, the same equation (4.11) is used with $m = 1$.

Since we do not use the cluster centers in the kernelized method, the algorithm should be rewritten using solely U and D_{ki}. There is also a problem of how to determine the initial values of D_{ki}. For this purpose we select c objects $y_1, \dots, y_c$ randomly from $\{x_1, \dots, x_N\}$ and let $W_i = y_i$ $(1 \le i \le c)$. Then

$$D_{ki} = \|\Phi(x_k) - \Phi(y_i)\|_H^2 = K(x_k, x_k) + K(y_i, y_i) - 2K(x_k, y_i). \tag{4.12}$$

Algorithm KFCM: *Kernelized FCM.*

KFCM1. Select randomly $y_1, \dots, y_c \in \{x_1, \dots, x_N\}$. Calculate D_{ki} by (4.12).
KFCM2. Calculate u_{ki} by (4.6), or if the entropy-based fuzzy c-means should be used, calculate u_{ki} by (4.7).

KFCM3. If the solution $U = (u_{ki})$ is convergent, stop. Else update D_{ki} using (4.11) (when the entropy-based fuzzy c-means should be used, update D_{ki} using (4.11) with $m = 1$). Go to KFCM2.
End KFCM.

It is not difficult to derive fuzzy classification functions. For the standard method, we have

$$D_i(x) = K(x,x) - \frac{2}{S_i(m)} \sum_{j=1}^{N} (u_{ji})^m K(x, x_j) + \frac{1}{S_i(m)^2} \sum_{j=1}^{N} \sum_{\ell=1}^{n} (u_{ji} u_{\ell i})^m K_{j\ell}, \tag{4.13}$$

$$\mathbf{U}_{\mathrm{kfcm}}^{(i)}(x) = \left[\sum_{j=1}^{c} \left(\frac{D_i(x)}{D_j(x)} \right)^{\frac{1}{m-1}} \right]^{-1} \tag{4.14}$$

For the entropy-based method, we calculate $D_i(x)$ by (4.13) with $m = 1$ and use

$$\mathbf{U}_{\mathrm{kefc}}^{(i)}(x) = \frac{\exp\left(-\frac{D_i(x)}{\nu}\right)}{\sum_{j=1}^{c} \exp\left(-\frac{D_j(x)}{\nu}\right)}. \tag{4.15}$$

Note 4.1.1. For deriving the solution in (4.8), let $\phi = (\phi_1, \dots, \phi_c)$ be an arbitrary element in H and ε be a small real number. From

$$\frac{J_{\mathrm{kfcm}}(U, W + \varepsilon\phi) - J_{\mathrm{kfcm}}(U, W)}{2} = \varepsilon \sum_{i=1}^{c} \sum_{k=1}^{N} (u_{ik})^m \langle \phi_i, \Phi(x_k) - W_i \rangle + o(\varepsilon^2),$$

The necessary condition for optimality is obtained by putting the term corresponding to the first order of ε to be zero:

$$\sum_{i=1}^{c} \sum_{k=1}^{N} (u_{ik})^m \langle \phi_i, \Phi(x_k) - W_i \rangle = 0.$$

Since ϕ_i is arbitrary, we have the condition for optimality:

$$\sum_{k=1}^{N} (u_{ik})^m (\Phi(x_k) - W_i) = 0.$$

4.1.2 Kernelized Crisp c-Means Algorithm

A basic algorithm of the kernelized crisp c-means clustering is immediately derived from **KFCM**. Namely, the same procedure should be used with different formulas.

Algorithm KCCM: *Kernelized Crisp c-Means.*

KCCM1. Select randomly $y_1, \dots, y_c \in \{x_1, \dots, x_N\}$. Calculate D_{ki} by (4.12).
KCCM2. Calculate u_{ki} to the cluster of the nearest center:

$$u_{ki} = \begin{cases} 1 & (i = \arg\min_{1 \le j \le c} D_{kj}), \\ 0 & (\text{otherwise}). \end{cases}$$

KCCM3. If the solution $U = (u_{ki})$ is convergent, stop. Else update D_{ki}:

$$D_{ki} = K_{kk} - \frac{2}{|G_i|} \sum_{x_j \in G_i} K_{jk} + \frac{1}{|G_i|^2} \sum_{x_j \in G_i} \sum_{x_\ell \in G_i} K_{j\ell}. \tag{4.16}$$

where $G_i = \{\, x_k : u_{ki} = 1,\ k = 1, \dots, N \,\}$. Go to KCCM2.
End KCCM.

The derivation of a sequential algorithm is also possible [107]. In the following we first state this algorithm and then describe how the value of the objective function and the distances are updated.

Algorithm KSCCM: *Kernelized Sequential Crisp c-Means.*

KSCCM1. Take c points y_j $(1 \le j \le c)$ randomly from X and use them as initial cluster centers W_i $(W_i = y_i)$. Calculate

$$D_{kj} = K(x_k, x_k) - 2K(x_k, y_j) + K(y_j, y_j).$$

KSCCM2. Repeat the next KSCCM3 until the decrease of the value of the objective function becomes negligible.
KSCCM3. Repeat KSCCM3a and KSCCM3b for $\ell = 1, \dots, n$.
KSCCM3a. For $x_\ell \in G_i$, calculate

$$j = \arg\min_{1 \le r \le c} D_{\ell r}.$$

using (4.16).
KSCCM3b If $i \ne j$, reallocate x_ℓ to G_j:

$$G_j = G_j \cup \{x_\ell\}, \quad G_i = G_i - \{x_\ell\}.$$

Update $\|\Phi(x_k) - W_i\|^2$ for $x_k \in G_i$ and $\|\Phi(x_{k'}) - W_j\|^2$ for $x_{k'} \in G_j$ and update the value of the objective function. Notice that W_i and W_j change.
End KSCCM.

We next describe how the quantities in KSCCM3b are updated. For this purpose let

$$J = J_{\mathrm{cm}}(\boldsymbol{G}, V) = \sum_{i=1}^{c} \sum_{x_k \in G_i} D(x_k, v_i)$$

be an abbreviated symbol for the objective function defined by (2.7). We also put

$$J_i = \sum_{x_k \in G_i} D_{ki} = \sum_{x_k \in G_i} \|\Phi(x_k) - W_i\|,$$

whence

$$J = \sum_{i=1}^{c} J_i.$$

Assume that the object x_ℓ moves from G_j to G_h. Put

$$G_j^* = G_j - \{x_\ell\}, \quad G_h^* = G_h \cup \{x_\ell\}.$$

Suppose $\bar{J}_j$, $\bar{J}_h$, $\bar{J}$ are the values of the objective functions before x_ℓ moves; J_j^*, J_h^*, J^* are the values values after x_ℓ has moved; W_j^* and W_h^* are cluster centers after x_ℓ has moved, while W_j and W_h are those before x_ℓ moves. Then

$$J^* = \bar{J} - \bar{J}_j - \bar{J}_h + J_j^* + J_h^*.$$

We have

$$\begin{aligned}
J_j^* &= \sum_{x \in G_j} \|\Phi(x) - W_j^*\|^2 + \|\Phi(x_\ell) - W_j^*\|^2 \\
&= \sum_{x \in G_j} \|\Phi(x) - W_j - \frac{\Phi(x_\ell) - W_j}{N_j + 1}\|^2 + \|\frac{N_j}{N_j + 1}(\Phi(x_\ell) - W_j)\|^2 \\
&= J_j + \frac{N_j}{N_j + 1}\|(\Phi(x_\ell) - W_j)\|^2 \\
&= J_j + \frac{N_j}{N_j + 1} D_{\ell j}
\end{aligned}$$

$$\begin{aligned}
J_h^* &= \sum_{x \in G_j} \|\Phi(x) - W_h^*\|^2 - \|\Phi(x_\ell) - W_h^*\|^2 \\
&= \sum_{x \in G_j} \|\Phi(x) - W_h + \frac{\Phi(x_\ell) - W_h}{N_h - 1}\|^2 - \|\frac{N_h}{N_h - 1}(\Phi(x_\ell) - W_h)\|^2 \\
&= J_h - \frac{N_h}{N_h - 1}\|(\Phi(x_\ell) - W_h)\|^2 \\
&= J_h - \frac{N_h}{N_h - 1} D_{\ell h}
\end{aligned}$$

in which $N_j = |G_j|$. Notice that $D_{\ell i} = \|\Phi(x_\ell) - W_i\|$ is calculated by (4.16). We thus have algorithm **KSCCM**.

4.1.3 Kernelized Learning Vector Quantization Algorithm

Let us remind the algorithm **LVQC** in which the updating equations for the cluster centers are

$$\begin{aligned}
m_l(t+1) &= m_l(t) + \alpha(t)[x(t) - m_l(t)], \\
m_i(t+1) &= m_i(t), \quad i \neq l, \\
t &= 1, 2, \ldots
\end{aligned}$$

and consider the kernelization of **LVQC** [70]. As usual, the center should be eliminated and updating distances alone should be employed in the algorithm. We therefore put

$$D_{kl}(t) = \|\Phi(x_k) - W_l(t)\|_H^2 \tag{4.17}$$

and express $D_{kl}(t+1)$ in terms of distances $D_{kj}(t)$, $j = 1, \dots, c$.

For simplicity, put $\alpha = \alpha(t)$, then

$$\begin{aligned}
&D_{ki}(t+1) - (1-\alpha)D_{ki}(t) + \alpha(1-\alpha)D_{li}(t) \\
&= K_{kk} + \alpha^2 K_{ll} - (1-\alpha)K_{kk} + \alpha(1-\alpha)K_{ll} - 2\alpha K_{kl} \\
&\quad + \{(1-\alpha)^2 - (1-\alpha) + \alpha(1-\alpha)\}\|m_i(t)\|^2 \\
&= \alpha(K_{kk} - 2K_{kl} + K_{ll})
\end{aligned}$$

where

$$K_{kk} = K(x_k, x_k), \quad K_{kl} = K(x_k, x_l).$$

Hence the next equation holds.

$$D_{ki}(t+1) = (1-\alpha)D_{ki}(t) - \alpha(1-\alpha)D_{li}(t) + \alpha(K_{kk} - 2K_{kl} + K_{ll}). \tag{4.18}$$

We thus have the next algorithm.

Algorithm KLVQC: *Kernelized LVQ Clustering.*

KLVQC1 Determine initial values of D_{ki}, $i = 1, \dots, c$, $k = 1, \dots, N$, by randomly taking c objects $y_1, \dots, y_c$ from $x_1, \dots, x_N$ and set them to be cluster centers.

KLVQC2 Find

$$D_{kl}(t) = \arg \min_{1 \le i \le c} D_{ki}(t)$$

and put x_k into G_l.

KLVQC3 Update $D_{ik}(t)$ by (4.18). Go to KLVQC2.

End KLVQC.

4.1.4 An Illustrative Example

A typical example is given to show how the kernelized methods work to produce clusters with nonlinear boundaries. Figure 4.1 shows such an example of a circular cluster inside another group of objects of a ring shape. Since this data set is typical for discussing capability of kernelized methods, we call it a 'circle and ring' data.

The question here is whether we can separate these two groups. Obviously, the methods of crisp and fuzzy c-means cannot, since they have linear cluster boundaries. A way to separate these two groups using the noise clustering is possible by assuming the outer group to be in the noise cluster, but this is rather an *ad hoc* method and far from a genuine method for the separation.

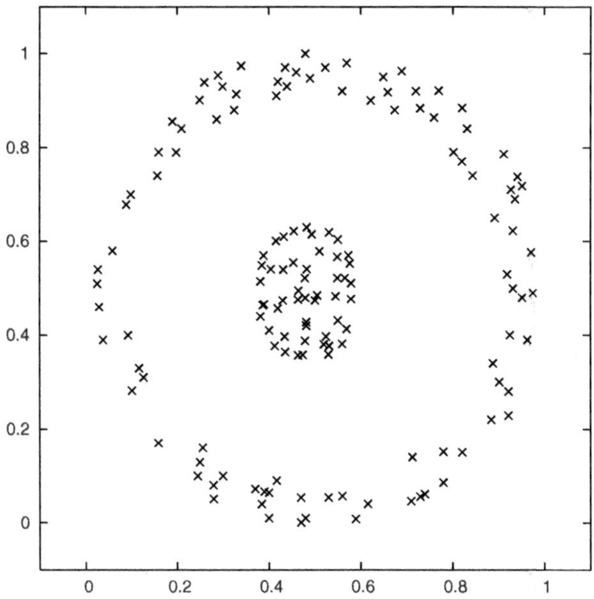

Fig. 4.1. An example of a circle and a ring around the circle

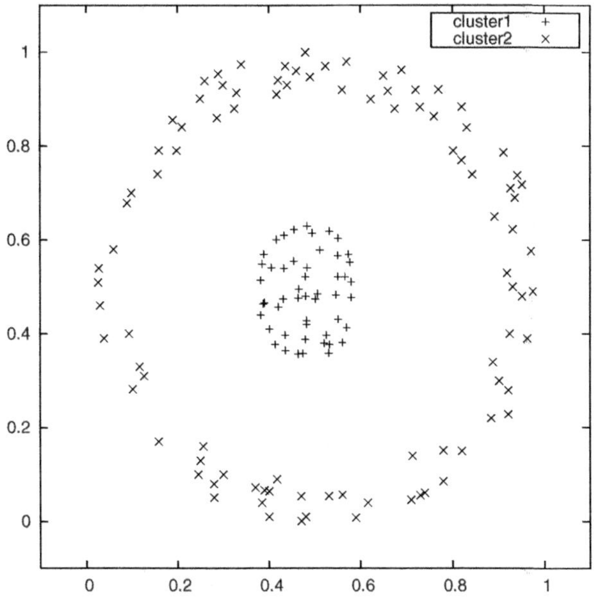

Fig. 4.2. Two clusters from **KCCM** and **KLVQC** ($c = 2$, $\lambda = 20$)

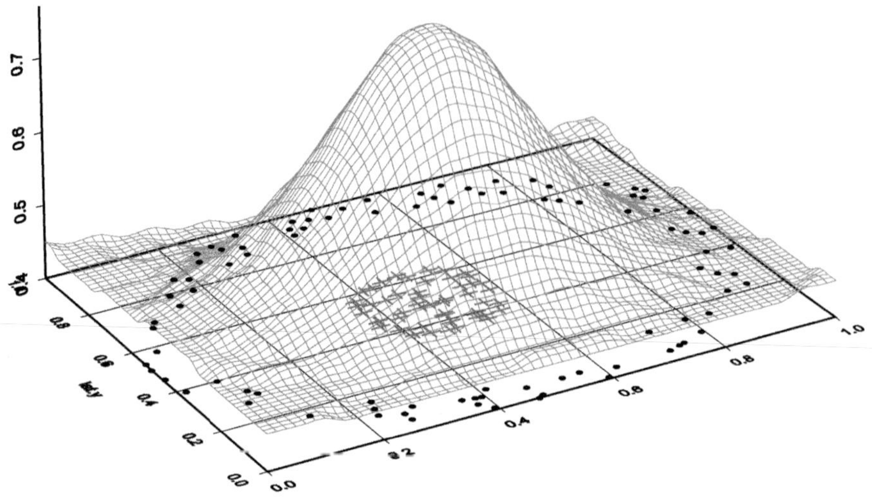

Fig. 4.3. Two clusters and a classification function from a 'circle and ring' data; standard fuzzy c-means with $m = 2$ is used

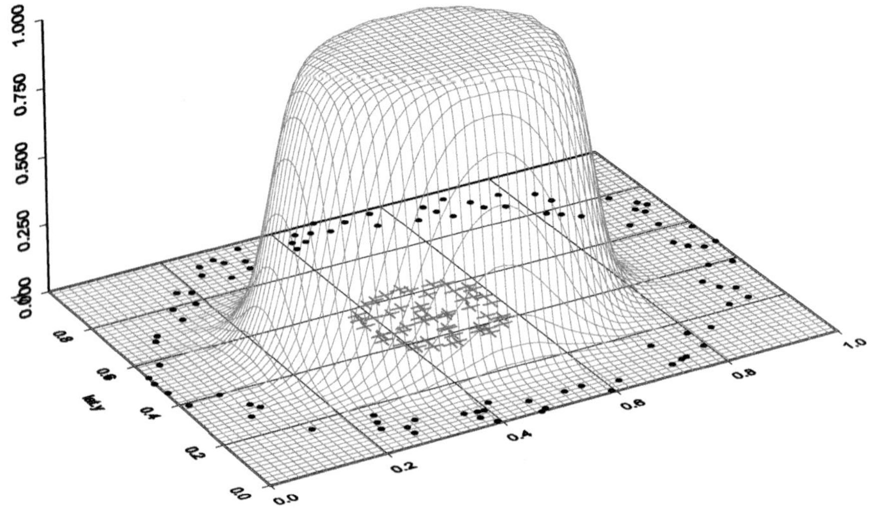

Fig. 4.4. Two clusters and a classification function from the 'circle and ring' data; entropy-based fuzzy c-means with $\nu^{-1} = 10.0$ is used

Figure 4.2 shows the output of two clusters from **KCCM**; the same output has also been obtained from **KSCCM** and **KLVQC**. We observe that the two groups are separated perfectly.

Figures 4.3 and 4.4 show classification functions of the inner cluster by the methods of the standard and the entropy-based fuzzy c-means, respectively. As the number of clusters are two ($c = 2$) and the sum of the two classification functions are equal to unity:

$$\sum_{i=1}^{c} \mathbf{U}^{(i)}(x; V) = 1,$$

one can see that this classification function implies successful separation of the two groups. Notice also the difference between the two classification functions in the two figures produced by the different algorithms. When we crisply reallocate the objects by the maximum membership rule (2.55), we have the same clusters as those in Figure 4.2.

There are many variations and applications of the kernelized methods of clustering, some of which will be discussed later.

4.2 Similarity Measure in Fuzzy c-Means

Up to now, we have assumed the squared Euclidean distance $D_{ki} = \|x_k - v_i\|^2$ as the dissimilarity measure between an object and a cluster center. There are, however, many other similarity and dissimilarity measures. For example, the Manhattan distance which is also called the city-block or L_1 distance is another important dissimilarity measure, and we will study algorithms based on the L_1 distance later.

Another class consists of similarity measures instead of dissimilarity: a similarity measure between arbitrary pair $x, x' \in X = \{x_1, \dots, x_N\}$ is denoted by $S(x, x')$ which takes a real value, which is symmetric with respect to the two arguments:

$$S(x, x') = S(x', x), \qquad \forall x, x' \in X. \tag{4.19}$$

In contrast to a dissimilarity measure, a large value of $S(x, x')$ means x and x' are near, while a small value of $S(x, x')$ means x and x' are distant. In particular, we can assume

$$S(x, x) = \max_{x' \in X} S(x, x'). \tag{4.20}$$

In the basic framework that does not accept an *ad hoc* technique, we do not study all similarity measures having the above properties. Instead, the discussion is limited to a well-known and useful measure of the cosine correlation that has frequently been discussed in information retrieval and document clustering [142, 161, 105].

Assume the inner product of the Euclidean space $\boldsymbol{R}^p$ be

$$\langle x, y \rangle = x^\top y = y^\top x = \sum_{j=1}^{p} x^j y^j,$$

then the cosine correlation is defined by

$$S_{\cos}(x, y) = \frac{\langle x, y \rangle}{\|x\| \|y\|}. \tag{4.21}$$

The name of *cosine* correlation comes from the simple fact that if we denote the angle between the two vectors x and y by $\theta(x, y)$, then

$$S_{\cos}(x, y) = \cos\theta(x, y).$$

holds. We immediately notice $0 \leq S_{\cos}(x, y) \leq 1$.

We now consider c-means clustering using the cosine correlation. For simplicity we omit the subscript of 'cos' and write $S(x, y)$ instead of $S_{\cos}(x, y)$, as we solely use the cosine correlation in this section.

A natural idea is to define a dissimilarity from $S(x, y)$. We hence put

$$D(x, v) = 1 - S(x, v).$$

Then $D(x, v)$ satisfies all properties required for a dissimilarity measure. We define the objective function for crisp c-means clustering:

$$J'(U, V) = \sum_{i=1}^{c} \sum_{k=1}^{N} u_{ki} D(x_k, v_i).$$

We immediately have

$$J'(U, V) = N - \sum_{i=1}^{c} \sum_{k=1}^{N} u_{ki} S(x_k, v_i).$$

This means that we should directly handle the measure $S_{\cos}(x, y)$. We therefore define

$$J(U, V) = \sum_{i=1}^{c} \sum_{k=1}^{N} u_{ki} S(x_k, v_i). \tag{4.22}$$

Accordingly, the algorithm should be alternate maximization instead of minimization: iteration of FCM2 now is

$$\bar{U} = \arg \max_{U \in \mathcal{U}_f} J(U, \bar{V}), \tag{4.23}$$

and FCM3 is

$$\bar{V} = \arg\max_{V} J(\bar{U}, V). \tag{4.24}$$

It is easy to see that

$$\bar{u}_{ki} = \begin{cases} 1 & (i = \arg \max\limits_{1 \leq j \leq c} S(x_k, v_j)), \\ 0 & (\text{otherwise}). \end{cases} \tag{4.25}$$

In order to derive the solution of $\bar{V}$, consider the next problem:

$$\text{maximize} \sum_{k=1}^{N} u_{ki} \frac{\langle x_k, v_i \rangle}{\|x_k\|}$$

$$\text{subject to } \|v_i\|^2 = 1.$$

Put

$$L(v_i,\gamma)=\sum_{k=1}^{N}u_{ki}\frac{\langle x_k,v_i\rangle}{\|x_k\|}+\gamma(\|v_i\|^2-1)$$

with the Lagrange multiplier γ, we have

$$\sum_{k=1}^{N}u_{ki}\frac{x_k}{\|x_k\|}+2\gamma v_i=0.$$

Using the constraint and eliminating γ, we obtain

$$\bar{v}_i=\frac{\displaystyle\sum_{k=1}^{N}\bar{u}_{ki}\frac{x_k}{\|x_k\|}}{\displaystyle\|\sum_{k=1}^{N}\bar{u}_{ki}\frac{x_k}{\|x_k\|}\|}. \tag{4.26}$$

We proceed to consider fuzzy c-means [112, 117]. For the entropy-based method, we define

$$J'(U,V)=\sum_{i=1}^{c}\sum_{k=1}^{N}u_{ki}D(x_k,v_i)+\nu\sum_{i=1}^{c}\sum_{k=1}^{N}u_{ki}\log u_{ki}. \tag{4.27}$$

$$J(U,V)=\sum_{i=1}^{c}\sum_{k=1}^{N}u_{ki}S(x_k,v_i)-\nu\sum_{i=1}^{c}\sum_{k=1}^{N}u_{ki}\log u_{ki}. \tag{4.28}$$

We immediately have

$$J'(U,V)+J(U,V)=N. \tag{4.29}$$

We therefore consider the alternate maximization of $J(U,V)$ by (4.28). In other words, iteration of FCM2 by (4.23) and FCM3 by (4.24).

It is immediate to see that the optimal solution $\bar{V}$ in FCM3 is given by the same equation (4.26). It is also easy to derive the optimal solution $\bar{U}$ in FCM2:

$$\bar{u}_{ki}=\frac{\exp\left(\frac{S(x_k,\bar{v}_i)}{\nu}\right)}{\displaystyle\sum_{j=1}^{c}\exp\left(\frac{S(x_k,\bar{v}_j)}{\nu}\right)}. \tag{4.30}$$

We next consider the objective function of the standard fuzzy c-means:

$$\begin{aligned}J'(U,V)&=\sum_{i=1}^{c}\sum_{k=1}^{N}(u_{ki})^mD(x_k,v_i),\\&=\sum_{i=1}^{c}\sum_{k=1}^{N}(u_{ki})^m(1-S(x_k,v_i))\end{aligned} \tag{4.31}$$

$$\check{J}(U,V)=\sum_{i=1}^{c}\sum_{k=1}^{N}(u_{ki})^mS(x_k,v_i). \tag{4.32}$$

We notice that a simple relation like (4.29) does not hold. The question hence is which to use either (4.31) or (4.32); the answer is that we cannot employ $\check{J}(U,V)$ with $S(x_k, v_i)$. The reason is clear: $\check{J}(U,V)$ is convex with respect to $(u_{ki})^m$ while it is concave with respect to $S(x_k, v_i)$. If we use $\check{J}(U,V)$, this means that we are trying to find a saddle point of the objective function, which is inappropriate for our purpose.

We thus employ $J'(U,V)$ in algorithm **FCM** and alternative minimization should be done as usual. The solution $\bar{U}$ in FCM2 is

$$\bar{u}_{ki} = \left[\sum_{j=1}^{c} \left(\frac{D(x_k, \bar{v}_i)}{D(x_k, \bar{v}_j)}\right)^{\frac{1}{m-1}}\right]^{-1} = \left[\sum_{j=1}^{c} \left(\frac{1 - S(x_k, \bar{v}_i)}{1 - S(x_k, \bar{v}_j)}\right)^{\frac{1}{m-1}}\right]^{-1}, \tag{4.33}$$

while $\bar{V}$ in **FCM3** is

$$\bar{v}_i = \frac{\displaystyle\sum_{k=1}^{N} (\bar{u}_{ki})^m \frac{x_k}{\|x_k\|}}{\displaystyle\|\sum_{k=1}^{N} (\bar{u}_{ki})^m \frac{x_k}{\|x_k\|}\|} . \tag{4.34}$$

4.2.1 Variable for Controlling Cluster Sizes

In section 3.2, we have introduced an additional variable $A = (\alpha_1, \dots, \alpha_c)$ to control cluster sizes (or cluster volumes) into the objective functions of fuzzy c-means and provided an extended algorithm **FCMA**. The application to the cosine correlation is immediate. For the crisp and entropy-based fuzzy c-means, we consider

$$J(U,V) = \sum_{i=1}^{c}\sum_{k=1}^{N} u_{ki} S(x_k, v_i) + \nu \log \alpha_i, \tag{4.35}$$

$$J(U,V,A) = \sum_{i=1}^{c}\sum_{k=1}^{N} u_{ki} S(x_k, v_i) - \nu \sum_{i=1}^{c}\sum_{k=1}^{N} u_{ki} \log \frac{u_{ki}}{\alpha_i}. \tag{4.36}$$

respectively, with the constraint $\mathcal{A} = \{A : \sum_{j=1}^{c} \alpha_j = 1, \alpha_i \geq 0, \text{all } i\}$. Hence the alternate maximization algorithm is used (we have omitted FCMA1 and FCMA5 in the following; see algorithm **FCMA** in section 3.2).

FCMA2. Calculate $\bar{U} = \arg\max_{U \in \mathcal{U}_f} J(U, \bar{V}, \bar{A})$.

FCMA3. Calculate $\bar{V} = \arg\max_{V} J(\bar{U}, V, \bar{A})$.

FCMA4. Calculate $\bar{A} = \arg\max_{A \in \mathcal{A}} J(\bar{U}, \bar{V}, A)$.

Optimal solutions are as follows.

Solution for FCMA2.
(i) crisp case:

$$\bar{u}_{ki} = \begin{cases} 1 & (i = \arg\max_{1 \le j \le c} \{S(x_k, v_j) + \nu \log \alpha_i\}), \\ 0 & (\text{otherwise}). \end{cases} \tag{4.37}$$

(ii) fuzzy case:

$$\bar{u}_{ki} = \frac{\alpha_i \exp\left(\frac{S(x_k,\bar{v}_i)}{\nu}\right)}{\sum_{j=1}^{c} \alpha_j \exp\left(\frac{S(x_k,\bar{v}_j)}{\nu}\right)}. \tag{4.38}$$

Solution for FCMA3. The same as (4.26): $\bar{v}_i = \dfrac{\sum_{k=1}^{N} \bar{u}_{ki} \dfrac{x_k}{\|x_k\|}}{\left\| \sum_{k=1}^{N} \bar{u}_{ki} \dfrac{x_k}{\|x_k\|} \right\|}$.

Solution for FCMA4. The same as (3.23): $\alpha_i = \sum_{k=1}^{N} u_{ki} \Big/ N$.

Classification function using the cosine correlation is immediately derived. For the entropy-based method with A, we have

$$\mathbf{U}^{(i)}_{\text{efca-cos}}(x; V) = \frac{\alpha_i \exp\left(\frac{S(x,\bar{v}_i)}{\nu}\right)}{\sum_{j=1}^{c} \alpha_j \exp\left(\frac{S(x,\bar{v}_j)}{\nu}\right)}. \tag{4.39}$$

Classification functions for other methods can directly be derived and we omit the detail.

The objective function and the solutions for the standard fuzzy c-means can also be derived: it is sufficient to note that the objective function is (3.16) with $D(x_k, v_i) = 1 - S(x_k, v_i)$ and the solutions for FCMA2, FCMA3, and FCMA4 are respectively given by (3.24), (4.26), and (3.26).

Note 4.2.1. In section 3.3, we considered covariance-like variables within clusters. To consider such a variable into the cosine correlation is difficult, since we should deal with a distribution on a surface of a unit hyper-sphere.

4.2.2 Kernelization Using Cosine Correlation

We proceed to consider kernelization of crisp and fuzzy c-means using cosine correlation [112, 117]. We employ the Gaussian kernel in applications.

Since objects $\Phi(x_1), \dots, \Phi(x_N)$ are in a high-dimensional Euclidean space H, the similarity should be the cosine correlation in H:

$$S_H(\Phi(x_k), W_i) = \frac{\langle \Phi(x_k), W_i \rangle_H}{\|\Phi(x_k)\|_H \|W_i\|_H}. \tag{4.40}$$

Hence we consider the next objective function to be maximized:

$$J(U, V) = \sum_{i=1}^{c} \sum_{k=1}^{N} u_{ki} S_H(\Phi(x_k), W_i) - \nu \sum_{i=1}^{c} \sum_{k=1}^{N} u_{ki} \log u_{ki}, \tag{4.41}$$

in which $\nu \geq 0$ and when $\nu = 0$, the function is for the crisp c-means clustering. Notice also that to include the additional variable A is straightforward but we omit it for simplicity.

Our purpose is to obtain an iterative algorithm such as the procedure **KFCM** in which W is eliminated and iteration of $\bar{U}$ and the distances D_{ki} is repeated. To this end we first notice the optimal solutions of U and W:

$$\bar{u}_{ki} = \frac{\exp\left(\dfrac{S_H(\Phi(x_k), \bar{W}_i)}{\nu}\right)}{\sum_{j=1}^{c} \exp\left(\dfrac{S_H(\Phi(x_k), \bar{W}_j)}{\nu}\right)}. \tag{4.42}$$

$$\bar{W}_i = \frac{\sum_{k=1}^{c} \bar{u}_{ki} \dfrac{\Phi(x_k)}{\|\Phi(x_k)\|}}{\left\| \sum_{k=1}^{c} \bar{u}_{ki} \dfrac{\Phi(x_k)}{\|\Phi(x_k)\|} \right\|}. \tag{4.43}$$

Notice $K_{j\ell} = K(x_j, x_\ell) = \langle \Phi(x_j), \Phi(x_\ell) \rangle$ and $\|W_i\|_H = 1$. Substituting (4.43) into (4.40) and after some manipulation, we obtain

$$S_H(\Phi(x_k), \bar{W}_i) = \frac{\sum_{\ell=1}^{N} \bar{u}_{\ell i} \dfrac{K_{k\ell}}{\sqrt{K_{\ell\ell}}}}{\sqrt{K_{kk} \sum_{j,\ell=1}^{N} \bar{u}_{ji} \bar{u}_{\ell i} \dfrac{K_{j\ell}}{\sqrt{K_{jj} K_{\ell\ell}}}}} \tag{4.44}$$

We now have the iterative procedure of **KFCM** using the cosine correlation in H. Take $y_1, \dots, y_c$ randomly from X and let $W_i = \Phi(y_i)$. Calculate initial values

$$S_H(\Phi(x_k), \bar{W}_i) = \frac{K(x_k, y_i)}{\sqrt{K(x_k, x_k) K(y_i, y_i)}}.$$

Then repeat (4.42) and (4.44) until convergence of $\bar{U}$.

In the crisp c-means (4.42) should be replaced by

$$\bar{u}_{ki} = \begin{cases} 1 & (i = \arg\max_{1\le j\le c} S_H(\Phi(x_k), \bar{W}_i)), \\ 0 & (\text{otherwise}), \end{cases} \tag{4.45}$$

and hence (4.45) and (4.44) should be repeated.

It should be noticed that when we employ the Gaussian kernel,

$$\|\Phi(x)\|^2 = K(x,x) = \exp(-\lambda\|x - x\|) = 1,$$

that is, $\|\Phi(x)\| = 1$ for all $x \in \boldsymbol{R}^p$. In such a case the above formula (4.44) is greatly simplified:

$$S_H(\Phi(x_k), \bar{W}_i) = \frac{\displaystyle\sum_{\ell=1}^{N} \bar{u}_{\ell i} K_{k\ell}}{\sqrt{\displaystyle\sum_{j,\ell=1}^{N} \bar{u}_{ji}\bar{u}_{\ell i} K_{j\ell}}}. \tag{4.46}$$

The procedure for the kernelized standard c-means can be derived in the same manner. We have

$$S_H(\Phi(x_k), \bar{W}_i) = \frac{\displaystyle\sum_{\ell=1}^{N} (\bar{u}_{\ell i})^m \frac{K_{k\ell}}{\sqrt{K_{\ell\ell}}}}{\sqrt{K_{kk} \displaystyle\sum_{j,\ell=1}^{N} (\bar{u}_{ji}\bar{u}_{\ell i})^m \frac{K_{j\ell}}{\sqrt{K_{jj}K_{\ell\ell}}}}} \tag{4.47}$$

and

$$\bar{u}_{ki} = \left[\sum_{j=1}^{c} \left(\frac{1 - S_H(\Phi(x_k), \bar{W}_i)}{1 - S_H(\Phi(x_k), \bar{W}_j)}\right)^{\frac{1}{m-1}}\right]^{-1}. \tag{4.48}$$

Hence (4.48) and (4.47) should be repeated.

Classification functions for the standard fuzzy c-means and the entropy-based fuzzy c-means use

$$S_H(\Phi(x), \bar{W}_i) = \frac{\displaystyle\sum_{\ell=1}^{N} (\bar{u}_{\ell i})^m \frac{K(x, x_\ell)}{\sqrt{K_{\ell\ell}}}}{\sqrt{K(x,x) \displaystyle\sum_{j,\ell=1}^{N} (\bar{u}_{ji}\bar{u}_{\ell i})^m \frac{K_{j\ell}}{\sqrt{K_{jj}K_{\ell\ell}}}}}, \tag{4.49}$$

$$\mathbf{U}^{(i)}_{\text{kfcm-cos}}(x) = \left[\sum_{j=1}^{c} \left(\frac{1 - S_H(\Phi(x), \bar{W}_i)}{1 - S_H(\Phi(x), \bar{W}_j)}\right)^{\frac{1}{m-1}}\right]^{-1}. \tag{4.50}$$

and

$$S_H(\Phi(x), \bar{W}_i) = \frac{\displaystyle\sum_{\ell=1}^{N} \bar{u}_{\ell i} \frac{K(x, x_\ell)}{\sqrt{K_{\ell\ell}}}}{\sqrt{K(x,x) \displaystyle\sum_{j,\ell=1}^{N} \bar{u}_{ji}\bar{u}_{\ell i} \frac{K_{j\ell}}{\sqrt{K_{jj}K_{\ell\ell}}}}}, \tag{4.51}$$

$$\mathbf{U}^{(i)}_{\text{kefc}-\cos}(x) = \frac{\exp\left(\dfrac{S_H(\Phi(x), \bar{W}_i)}{\nu}\right)}{\displaystyle\sum_{j=1}^{c} \exp\left(\dfrac{S_H(\Phi(x), \bar{W}_j)}{\nu}\right)}, \tag{4.52}$$

respectively.

4.2.3 Clustering by Kernelized Competitive Learning Using Cosine Correlation

We have studied a kernelized version of LVQ clustering algorithm earlier in this section. Here another algorithm of clustering based on competitive learning is given. Although this algorithm uses the scalar product instead of the Euclidean distance, the measure is the same as the cosine correlation, since all objects are normalized: $x_k \leftarrow x_k/\|x_k\|$.

We first describe this algorithm which is given in Duda *et al.* [30] and then consider its kernelization.

Algorithm CCL: *Clustering by Competitive Learning.*

CCL1. Normalize x_k: $x_k \leftarrow x_k/\|x_k\|$ $(k = 1, \dots, N)$ and randomly select cluster centers v_i $(i = 1, \dots, c)$ from $x_1, \dots, x_N$.
Set $t = 0$ and Repeat CCL2 and CCL3 until convergence.

CCL2. Select randomly x_k. Allocate x_k to the cluster i:

$$i = \arg \max_{1 \le j \le c} \langle x_k, v_i \rangle.$$

CCL3. Update the cluster center:

$$v_i \leftarrow v_i + \eta(t) x_k,$$
$$v_i \leftarrow v_i/\|v_i\|.$$

Let $t \leftarrow t + 1$.
End of CCL.

The parameter $\eta(t)$ should be taken to satisfy

$$\sum_{t=1}^{\infty} \eta(t) = 0, \quad \sum_{t=1}^{\infty} \eta^2(t) < \infty.$$

We proceed to consider kernelization of **CCL**. For this purpose let the reference vector, which is the cluster center, be W_i. Put

$$y_k \leftarrow \Phi(x_k)/\|\Phi(x_k)\|$$

and note that the allocation rule is

$$i = \arg \max_{1\le j\le c} \langle y_k, W_j\rangle.$$

Let

$$p(x_k, i; t) = \langle y_k, W_i\rangle$$

be the value of the scalar product at time t. From the definitions for updating:

$$v_i \leftarrow v_i + \eta(t)x_k, \quad v_i \leftarrow v_i/\|v_i\|,$$

we note

$$p(x_k, i; t+1) = \langle y_k, \frac{W_i + \eta(t)y_\ell}{\|W_i + \eta(t)y_\ell\|}\rangle.$$

where y_ℓ is the last object taken at time t. Put $V_i(t) = \|W_i\|^2$ and note that

$$\langle y_k, y_\ell\rangle = \langle \Phi(x_k)/\|\Phi(x_k)\|, \Phi(x_\ell)/\|\Phi(x_\ell)\|\rangle = \frac{K_{k\ell}}{\sqrt{K_{kk}K_{\ell\ell}}}.$$

We hence have

$$V_i(t+1) = V_i(t) + 2\eta(t)p(x_\ell, i; t) + \eta(t)K_{\ell\ell}, \tag{4.53}$$

$$p(x_k, i; t+1) = \frac{1}{\sqrt{V_i(t+1)}}\left(p(x_k, i; t) + \eta(t)\frac{K_{k\ell}}{\sqrt{K_{kk}K_{\ell\ell}}}\right). \tag{4.54}$$

We now have another kernelized clustering algorithm based on competitive learning which uses the scalar product.

Algorithm KCCL: *Kernelized Clustering by Competitive Learning.*

KCCL1. Select randomly c cluster centers $W_i = \Phi(x'_{j_i})$ from $(k = 1, \dots, N)$ and compute

$$p(x_k, i; 0) = K(x_k, x'_{j_i})\Big/\sqrt{K_{kk}K(x'_{j_i}, x'_{j_i})}$$

for $i = 1, \dots, c$ and $k = 1, \dots, N$.
Set $t = 0$ and Repeat KCCL2 and KCCL3 until convergence.

KCCL2. Select randomly x_k. Allocate x_k to the cluster i:

$$i = \arg \max_{1\le j\le c} p(x_k, j; t).$$

KCCL3. Update $p(x_k, i; t+1)$ using (4.53) and (4.54). Let $t \leftarrow t+1$.
End of KCCL.

Note 4.2.2. Many other studies have been done concerning application of kernel functions to clustering. For example we can consider kernelized fuzzy LVQ algorithms [115] and kernelized possibilistic clustering [116]. Since kernelized methods are recognized to be powerful and sometimes outperform traditional techniques, many variations of the above stated methods should further be studied.

4.3 Fuzzy c-Means Based on L_1 Metric

Another important class of dissimilarity is the L_1 metric:

$$D(x,y) = \|x-y\|_1 = \sum_{j=1}^{p} |x^j - y^j| \tag{4.55}$$

where $\|\cdot\|_1$ is the L_1 norm. The L_1 metric is also called the Manhattan distance or the city-block distance. The use of L_1 metric has been studied by several researchers [12, 73, 101, 103]. Bobrowski and Bezdek studied L_1 and L_∞ metrics and derived complicated algorithms; Jajuga proposed an iterative procedure of a fixed point type. We describe here a simple algorithm of low computational complexity based on the rigorous alternate minimization [101, 103].

The objective functions based on the L_1 metric are

$$\begin{aligned} J_{\text{fcm}}(U,V) &= \sum_{i=1}^{c}\sum_{k=1}^{N} (u_{ki})^m D(x_k, v_i) \\ &= \sum_{i=1}^{c}\sum_{k=1}^{N} (u_{ki})^m \|x_k - v_i\|_1 \end{aligned} \tag{4.56}$$

$$\begin{aligned} J_{\text{efc}}(U,V) &= \sum_{i=1}^{c}\sum_{k=1}^{N} u_{ki} D(x_k, v_i) + \nu \sum_{i=1}^{c}\sum_{k=1}^{N} u_{ki} \log u_{ki} \\ &= \sum_{i=1}^{c}\sum_{k=1}^{N} u_{ki} \|x_k - v_i\|_1 + \nu \sum_{i=1}^{c}\sum_{k=1}^{N} u_{ki} \log u_{ki}. \end{aligned} \tag{4.57}$$

It is easy to derive the solutions $\bar{U}$ for FCM2:

$$\bar{u}_{ki} = \left[\sum_{j=1}^{c} \left(\frac{\|x_k - \bar{v}_i\|_1}{\|x_k - \bar{v}_j\|_1} \right)^{\frac{1}{m-1}} \right]^{-1}, \tag{4.58}$$

$$\bar{u}_{ki} = \frac{\exp\left(-\dfrac{\|x_k - \bar{v}_i\|_1}{\nu} \right)}{\displaystyle\sum_{j=1}^{c} \exp\left(-\dfrac{\|x_k - \bar{v}_j\|_1}{\nu} \right)}, \tag{4.59}$$

respectively for the standard and entropy-based method.

The main problem is how to compute the cluster centers. In the following we mainly consider the standard fuzzy c-means, as it is simple to modify the algorithm to the case of the entropy-based fuzzy c-means. First note that

$$\begin{aligned} J_{\text{fcm}}(U,V) &= \sum_{i=1}^{c}\sum_{k=1}^{N} (u_{ki})^m \|x_k - v_i\|_1 \\ &= \sum_{i=1}^{c}\sum_{j=1}^{p}\sum_{k=1}^{N} (u_{ki})^m |x_k^j - v_i^j|. \end{aligned}$$

We put

$$F_{ij}(w) = \sum_{k=1}^{N} (u_{ki})^m |x_k^j - w|$$

as a function of a real variable w. Then

$$J_{\text{fcm}}(U, V) = \sum_{i=1}^{c} \sum_{j=1}^{p} F_{ij}(v_{ij}),$$

in which U does not represent variables but parameters.

To determine cluster centers, each $F_{ij}(w)$ $(1 \le i \le c,\ 1 \le j \le p)$ should be minimized with respect to the real variable w without any constraint. It is easily seen that the following properties are valid, of which proofs are omitted.

(A) $F_{ij}(w)$ is a convex [133] and piecewise affine function.

(B) The intersection between the set $X_j = \{x_1^j, ..., x_N^j\}$ and the set of the solutions of

$$\min_{w \in \boldsymbol{R}} F_{ij}(w) \tag{4.60}$$

is not empty.

In other words, at least one of the j-th coordinates of the points $x_1, \ldots, x_N$ is the optimal solution.

In view of property (B), we limit ourselves to the minimization problem

$$\min_{w \in X_j} F_{ij}(w) \tag{4.61}$$

instead of (4.60).

No simple formula for cluster centers in the L_1 case seems to be available, but an efficient algorithm of search for the solution of (4.61) should be considered using the above properties.

Two ideas are employed in the following algorithm: ordering of $\{x_k^j\}$ and derivative of F_{ij}. We assume that when $\{x_1^j, ..., x_N^j\}$ is ordered, subscripts are changed using a permutation function $q_j(k)$, $k = 1, \ldots, N$, that is, $x_{q_j(1)}^j \le x_{q_j(2)}^j \le \cdots \le x_{q_j(N)}^j$. Using $\{x_{q_j(k)}^j\}$,

$$F_{ij}(w) = \sum_{k=1}^{N} (u_{q_j(k)i})^m |w - x_{q_j(k)}^j|. \tag{4.62}$$

Although $F_{ij}(w)$ is not differentiable on $\boldsymbol{R}$, we extend the derivative of $F_{ij}(w)$ on $\{x_{q_j(k)}^j\}$:

$$dF_{ij}^+(w) = \sum_{k=1}^{N} (u_{q_j(k)i})^m \operatorname{sign}^+(w - x_{q_j(k)}^j) \tag{4.63}$$

where

$$\operatorname{sign}^+(z) = \begin{cases} 1 & (z \ge 0), \\ -1 & (z < 0). \end{cases}$$

Thus, $dF^+_{ij}(w)$ is a step function which is right continuous and monotone non-decreasing in view of its convexity and piecewise affine property. It now is easy to see that the minimizing element for (4.61) is one of $x^j_{q_j(k)}$ at which $dF^+_{ij}(w)$ changes its sign. More precisely, $x^j_{q_j(t)}$ is the optimal solution of (4.61) if and only if $dF^+_{ij}(w) < 0$ for $w < x^j_{q_j(t)}$ and $dF^+_{ij}(w) \geq 0$ for $w \geq x^j_{q_j(t)}$.

Let $w = x^j_{q_j(r)}$, then

$$dF^+_{ij}(x^j_{q_j(r)}) = \sum_{k=1}^{r}(u_{q_j(k)i})^m - \sum_{k=r+1}^{N}(u_{q_j(k)i})^m$$

These observations lead us to the next algorithm.

begin
  $S := -\sum_{k=1}^{N}(u_{ki})^m$;
  $r := 0$;
  while ($S < 0$) **do begin**
    $r := r + 1$;
    $S := S + 2(u_{q_j(r)i})^m$
  end;
  output $v^j_i = x^j_{q_j(r)}$ as the j-th coordinate of cluster center v_i
end.

It is easy to see that this algorithm correctly calculates the cluster center, the solution of (4.61).

This algorithm is a simple linear search on nodes of the piecewise affine function. It is very efficient, since at most $4n$ additions and n conditional branches should be processed. No multiplication is needed. It is unnecessary to calculate $(u_{ki})^m$ from u_{ki}, since it is sufficient to store $(u_{ki})^m$ itself.

Thus, the calculation of cluster centers in the L_1 case is simple and does not require much computation time, except that the ordering of $\{x^j_k\}$ for each coordinate j is necessary. Notice that the ordering is performed only once before the iteration of **FCM**. Further ordering is unnecessary during the iteration.

The computational complexity for calculating V in **FCM** is $O(np)$; the complexity of the ordering before the iteration is $O(np \log n)$.

We also notice that the algorithm for the entropy-based fuzzy c-means is directly obtained by putting $m = 1$ in the above algorithm.

4.3.1 Finite Termination Property of the L_1 Algorithm

Interestingly enough, the algorithm **FCM** using the L_1 metric can be proved to terminate after a finite number of iteration, like Proposition 2.3.1 in the crisp case. For the termination, however, the condition by which the algorithm is judged to be terminated should be the value of the objective function. For stating the next proposition, we note the current optimal solution in **FCM** be $(\bar{U}, \bar{V})$ and denote the last optimal solution be $(\hat{U}, \hat{V})$. We moreover denote $J_{\text{new}} = J(\bar{U}, \bar{V})$ and $J_{\text{prev}} = J(\hat{U}, \hat{V})$. Observe also the following two facts:

(a) In the alternate minimization, the value of the objective function is monotonically non-increasing.
(b) As described above, the optimal solutions of v_i^j takes the value on finite points in X_j.

Since V can take values on finite points, the values of the objective functions with the optimal solutions are also finite and non-increasing. Hence the objective function must take a stationary value after a finite number of iterations of the major iteration loop of **FCM**. We thus have the next proposition.

Proposition 4.3.1. *The algorithm* **FCM** *based on the L_1 metric (i.e., using (4.56) or (4.57) as objective functions) finally stops after a finite number of iterations of the major loop FCM2–FCM4 on the condition that the algorithm is judged to be terminated if the value of the objective function does not decrease:* $J_{\text{new}} = J_{\text{prev}}$.

4.3.2 Classification Functions in the L_1 Case

Apparently, the classification functions are

$$\mathbf{U}^{(i)}_{\text{fcmL1}}(x;V) = \left[\sum_{j=1}^{c}\left(\frac{\|x-v_i\|_1}{\|x-v_j\|_1}\right)^{\frac{1}{m-1}}\right]^{-1}, \tag{4.64}$$

$$\mathbf{U}^{(i)}_{\text{efmL1}}(x;V) = \frac{\exp\left(-\dfrac{\|x-v_i\|_1}{\nu}\right)}{\sum_{j=1}^{c}\exp\left(-\dfrac{\|x-v_j\|_1}{\nu}\right)}. \tag{4.65}$$

As in section 2.7, we can analyze theoretical properties of these functions. Let us first consider $\mathbf{U}^{(i)}_{\text{fcmL1}}(x;V)$. From

$$1/\mathbf{U}^{(i)}_{\text{fcmL1}}(x;V) - 1 = \sum_{j\neq i}\left(\frac{\|x-v_i\|_1}{\|x-v_j\|_1}\right)^{\frac{1}{m-1}}, \tag{4.66}$$

we have

$$1/\mathbf{U}^{(i)}_{\text{fcm}}(x;V) - 1 \to 0 \quad \text{as} \quad x \to v_i.$$

Noting that $\mathbf{U}^{(i)}_{\text{fcmL1}}(x;V)$ is continuous whenever $x \neq v_i$, we have proved $\mathbf{U}^{(i)}_{\text{fcmL1}}$ $(x;V)$ is continuous on $\boldsymbol{R}^p$. Since it is easy to see that $\mathbf{U}^{(i)}_{\text{efcL1}}(x;V)$ is continuous everywhere, we have

Proposition 4.3.2. $\mathbf{U}^{(i)}_{\text{fcmL1}}(x;V)$ *and* $\mathbf{U}^{(i)}_{\text{efcL1}}(x;V)$ *are continuous on* $\boldsymbol{R}^p$.

It is moreover easy to see that

$$1/\mathbf{U}^{(i)}_{\text{fcmL1}}(x;V) - 1 \to c-1 \quad \text{as} \quad \|x\|_1 \to \infty,$$

and

$$1/\mathbf{U}^{(i)}_{\mathrm{fcmL1}}(x;V) > 1, \quad \forall x \neq v_i.$$

Hence we obtain

Proposition 4.3.3. *The function* $\mathbf{U}^{(i)}_{\mathrm{fcmL1}}(x;V)$ *takes its maximum value* 1 *at* $x = v_i$ *while it tends to* $1/c$ *as* $\|x\| \to +\infty$*:*

$$\max_{x \in \boldsymbol{R}^p} \mathbf{U}^{(i)}_{\mathrm{fcmL1}}(x;V) = \mathbf{U}^{(i)}_{\mathrm{fcmL1}}(v_i;V) = 1 \tag{4.67}$$

$$\lim_{\|x\| \to +\infty} \mathbf{U}^{(i)}_{\mathrm{fcmL1}}(x;V) = \frac{1}{c}. \tag{4.68}$$

The classification function $\mathbf{U}^{(i)}_{\mathrm{efcL1}}(x;V)$ for the entropy-based method has more complicated properties and we omit the detail.

4.3.3 Boundary between Two Clusters in the L_1 Case

As in the usual dissimilarity of the squared Euclidean distance, frequently the crisp reallocation has to be done by the maximum membership rule:

$$x_k \in G_i \iff u_{ki} = \max_{1 \le j \le c} u_{kj}.$$

Furthermore, the maximum membership rule is applied to the classification functions:

$$x \in G_i \iff \mathbf{U}^{(i)}_{\mathrm{fcmL1}}(x;V) > \mathbf{U}^{(j)}_{\mathrm{fcmL1}}(x;V), \quad \forall j \neq i,$$

or

$$x \in G_i \iff \mathbf{U}^{(i)}_{\mathrm{efcL1}}(x;V) > \mathbf{U}^{(j)}_{\mathrm{efcL1}}(x;V), \quad \forall j \neq i.$$

In the both case, the allocation rules are reduced to

$$x \in G_i \iff \|x - v_i\|_1 < \|x - v_j\|_1, \quad \forall j \neq i. \tag{4.69}$$

Thus the crisp reallocation uses the nearest center rule (4.69) for the both methods. Notice that we use the L_1 distance now.

This means that we should investigate the shape of boundary between two centers:

$$B_{L_1}(v,v') = \{x \in \boldsymbol{R}^p : \|x - v\|_1 = \|x - v'\|_1\}.$$

For simplicity we consider a plane ($p = 2$). For the Euclidean distance the boundary is the line intersects vertically the segment connecting v and v' at the midpoint.

In the L_1 plane, the boundary has a piecewise linear shape. Figure 4.5 illustrates an example of the boundary that consists of three linear segments on a plane, where v and v' are shown by $\times$. The dotted segments and dotted inner square provide supplementary information. The outer square implies the region of the plane for the illustration.

Let us observe the dotted segment connecting v and v'. The boundary intersects this segment at the midpoint inside the dotted square which includes the

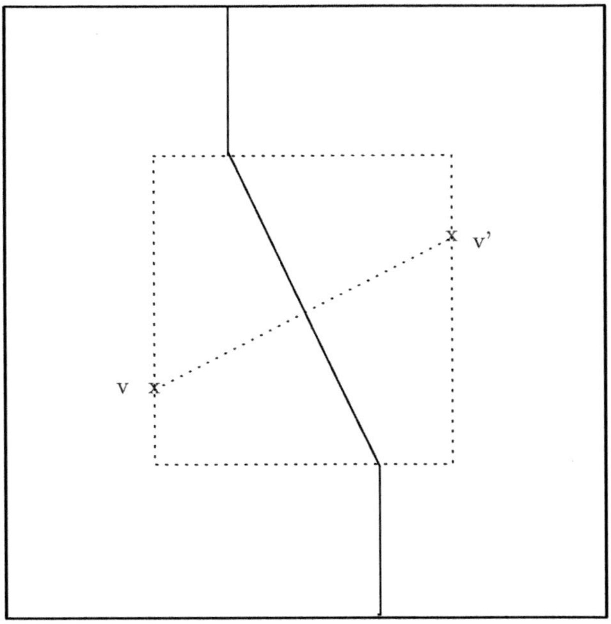

Fig. 4.5. An example of equidistant lines from the two points shown by **x** on a plane where the L_1 metric is assumed

two points v and v'. Outside of the dotted square the boundary is parallel to an axis. In this figure the lines are parallel to the vertical axis, since the boundary meets the upper and lower edges of the dotted square; if the boundary meets the right and left edges of the dotted square (which means that the two centers are in the upper and lower edges of the square), the boundary will be parallel to the horizontal axis.

Thus the Voronoi sets based on the L_1 metric consist of such a complicated combination of line segments, even when $p = 2$.

4.4 Fuzzy c-Regression Models Based on Absolute Deviation

In section 3.7, we have studied fuzzy c-regression models in which the sum of errors is measured by the least square. Another method for regression models is based on the least absolute deviation which requires more computation than the least square but is known to have robustness [11]. In this section we consider fuzzy c-regression models based on least absolute deviation [11, 73]. Notice that when we compare the least square to the squared Euclidean distance, the least absolute deviation described below can be compared to the L_1 metric.

Let us remind that what we need to have is the c-regression models:

$$y = f_i(x; \beta_i) + e_i, \quad i = 1, \ldots, c.$$

In the case of the least square, we minimize the sum of squared error e_i^2:

$$D_{ki} = (y_k - f_i(x_k; \beta_i))^2.$$

In contrast, we minimize the absolute value of error $|e_i|$:

$$D_{ki} = |y_k - f_i(x_k; \beta_i)|.$$

Specifically, we assume the linear regression models as in section 3.7:

$$f_i(x; \beta_i) = \sum_{j=1}^{p} \beta_i^j x^j + \beta_i^{p+1}.$$

and put

$$\begin{aligned} z &= (x, 1)^\top = (x^1, \ldots, x^p, 1)^\top, \\ z_k &= (x_k, 1)^\top = (x_k^1, \ldots, x_k^p, 1)^\top, \\ \beta_i &= (\beta_i^1, \ldots, \beta_i^{p+1}), \end{aligned}$$

whence we define

$$D_{ki} = D((x_k, y_k), \beta_i) = |y_k - \sum_{j=1}^{p} \beta_i^j x^j + \beta_i^{p+1}| = |y_k - \langle z_k, \beta_i \rangle|. \tag{4.70}$$

Using D_{ki} by (4.70), the next two objective functions are considered for algorithm **FCRM**.

$$\begin{aligned} J_{\text{fcrmabs}}(U, B) &= \sum_{i=1}^{c} \sum_{k=1}^{N} (u_{ki})^m D((x_k, y_k), \beta_i) \\ &= \sum_{i=1}^{c} \sum_{k=1}^{N} (u_{ki})^m D_{ki}, \end{aligned} \tag{4.71}$$

$$\begin{aligned} J_{\text{efcrmabs}}(U, B) &= \sum_{i=1}^{c} \sum_{k=1}^{N} \{u_{ki} D((x_k, y_k), \beta_i) + \nu u_{ki} \log u_{ki}\} \\ &= \sum_{i=1}^{c} \sum_{k=1}^{N} \{u_{ki} D_{ki} + \nu u_{ki} \log u_{ki}\} \end{aligned} \tag{4.72}$$

where $B = (\beta_1, \ldots, \beta_c)$. Notice that (4.71) and (4.72) appear the same as (3.64) and (3.65), respectively, but the dissimilarity D_{ki} is different.

The solutions for the optimal U are

$$\bar{u}_{ki} = \left[\sum_{j=1}^{c} \left(\frac{D_{ki}}{D_{kj}} \right)^{\frac{1}{m-1}} \right]^{-1}, \tag{4.73}$$

$$\bar{u}_{ki} = \frac{\exp\left(-\frac{D_{ki}}{\nu}\right)}{\sum_{j=1}^{c} \exp\left(-\frac{D_{kj}}{\nu}\right)}, \tag{4.74}$$

for $J = J_{\text{fcrmabs}}$ and $J = J_{\text{efcrmabs}}$, respectively.

On the other hand, the optimal B requires a more complicated algorithm than the least square. Let

$$J^i(\beta_i) = \sum_{k=1}^{N} (u_{ki})^m \big| y_k - \langle \beta_i, z_k \rangle \big|, \tag{4.75}$$

regarding u_{ki} as parameters. Since

$$J_{\text{fcrmabs}}(U, B) = \sum_{i=1}^{c} J^i(\beta_i),$$

each $J^i(\beta_i)$ should be indenpendently minimized in order to obtain the optimal B. The minimization of $J^i(\beta_i)$ is reduced to the next linear programming problem [11].

$$\min \sum_{k=1}^{N} (u_{ki})^m r_{ki} \tag{4.76}$$

$$y_k - \langle \beta_i, z_k \rangle \leq r_{ki} \tag{4.77}$$

$$y_k - \langle \beta_i, z_k \rangle \geq -r_{ki} \tag{4.78}$$

$$r_{ik} \geq 0, \quad k = 1, \ldots, N,$$

where the variables are β_i and r_{ki}, $k = 1, \ldots, N$.

To observe the minimization of $J^i(\beta_i)$ is equivalent to this linear programming problem, note $|y_k - \langle \beta_i, z_k \rangle| \leq r_{ki}$ from (4.77) and (4.78). Suppose $u_{ki} > 0$, the minimization of (4.76) excludes $|y_k - \langle \beta_i, z_k \rangle| < r_{ki}$ and $|y_k - \langle \beta_i, z_k \rangle| = r_{ki}$ holds. If $u_{ik} = 1$ and $u_{jk} = 0$ $(j \neq i)$, then the corresponding r_{kj} does not affect the minimization of (4.76), hence we can remove r_{kj} and the corresponding constraint from the problem. Thus the equivalence of the both problems is valid. Hence we have the solution β_i by solving the problem (4.76)–(4.78).

4.4.1 Termination of Algorithm Based on Least Absolute Deviation

The fuzzy regression models based on the least absolute deviation terminates after a finite number of iterations of the major loop of **FCRM**, as in the case of the L_1 metric in **FCM**. Namely we have the next proposition.

Proposition 4.4.1. *The algorithm* **FCRM** *using the objective function* $J = J_{\text{fcrmabs}}$ *or* $J = J_{\text{efcrmabs}}$ *terminates after a finite number of iterations of the major loop in* **FCRM**, *provided that the convergence test uses the value of the objective function:* $J_{\text{new}} = J_{\text{prev}}$.

Note 4.4.1. As in Proposition 4.3.1, J_{new} means the current optimal value of the objective function, while J_{prev} is the last optimal value. Hence $J_{\text{new}} \leq J_{\text{prev}}$.

The rest of this section is devoted to the proof of this proposition. Let

$$\mathsf{X} = \begin{pmatrix} x_1^1 & x_1^2 & \cdots & x_1^p & 1 \\ x_2^1 & x_2^2 & \cdots & x_2^p & 1 \\ \cdot & \cdot & \cdots & \cdot & \cdot \\ x_N^1 & x_N^2 & \cdots & x_N^p & 1 \end{pmatrix},$$

and assume $\operatorname{rank} \mathsf{X} = p+1$. Put $\mathsf{Y} = (y_1, \dots, y_N)^\top$. Assume that $w_1, w_2, \dots, w_N$ are positive constants and we consider minimization of

$$F(\beta) = \sum_{k=1}^{N} w_k |y_k - \langle \beta, z_k \rangle| \tag{4.79}$$

using the variable $\beta = (\beta^1, \dots, \beta^{p+1})^\top$ instead of (4.75) for simplicity. Suppose $V = \{i_1, \dots, i_\ell\}$ is a subset of $\{1, 2, \dots, N\}$ and

$$\mathsf{X} = \begin{pmatrix} x_1^1 & x_1^2 & \cdots & x_1^p & 1 \\ x_2^1 & x_2^2 & \cdots & x_2^p & 1 \\ \cdot & \cdot & \cdots & \cdot & \cdot \\ x_N^1 & x_N^2 & \cdots & x_N^p & 1 \end{pmatrix},$$

the matrix corresponding to V. Put also $\mathsf{Y}(V) = (y_{i_1}, \dots, y_{i_\ell})^\top$. We now have the next lemma.

Lemma 4.4.1. *The minimizing solution for* $F(\beta)$ *is obtained by taking a subset* $Z = \{i_1, \dots, i_{p+1}\}$ *of* $\{1, 2, \dots, N\}$ *and solving*

$$\mathsf{Y}(Z) = \mathsf{X}(Z)\beta.$$

Since the number of elements in Z *is* $p+1$, *this means that by taking* $p+1$ *subset of data* $\{(x_k, y_k)\}$ *and fitting* $y = \langle \beta, z \rangle$, *we can minimize (4.79).*

In order to prove Lemma 4.4.1, we will prove the next lemma.

Lemma 4.4.2. *For the subset* $V = \{i_1, \dots, i_q\}$ *of* $\{1, 2, \dots, N\}$, *let an arbitrary solution of* $\mathsf{Y}(V) = \mathsf{X}(V)\beta$ *be* $\beta(V)$. *If* $|V| = q < p+1$ *(*$|V|$ *is the number of elements in* V*), then there exists* $j \in \{1, \dots, N\} - V$ *such that the solution* $\beta(V')$ *of* $\mathsf{Y}(V') = \mathsf{X}(V')\beta$ *where* $V' = V \cup \{j\}$ *decreases the value of the objective function. (If* $V = \emptyset$, *then we can choose an arbitrary* $\beta(V) \in \boldsymbol{R}^{p+1}$.*)*

Proof. Let us prove Lemma 4.4.2. For simplicity we write $\beta = \beta(V)$ and assume $V = \{k : y_k = \langle \beta, z_k \rangle, 1 \leq k \leq N\}$, $|V| = q < p+1$. Since $q < \text{rank}\,(\mathsf{X})$, there exists $\rho \in \boldsymbol{R}^{p+1}$ such that

$$\mathsf{X}(V)\rho = 0, \qquad \mathsf{X}\rho \neq 0.$$

Hence

$$\langle z_k, \rho \rangle = 0, \quad \forall k \in V,$$
$$\langle z_i, \rho \rangle \neq 0, \quad \exists i \notin V$$

holds. Put

$$\gamma(t) = \beta + t\rho, \quad t \in \boldsymbol{R}.$$

Then,

$$F(\gamma(t)) = \sum_{j \notin V} w_j |y_j - \langle \beta, z_j \rangle - t\langle \rho, z_j \rangle|.$$

Put $\eta_j = y_j - \langle \beta, z_j \rangle (\neq 0)$ and $\zeta_j = \langle \rho, z_j \rangle$ for simplicity.

$$F(\gamma(t)) = \sum_{j \notin V} w_j |\eta_j - t\zeta_j|.$$

From the assumption there exists nonzero ζ_j. Hence if we put $V'' = \{j : \zeta_j \neq 0\}$, $V'' \neq \emptyset$ and we have

$$F(\gamma(t)) = \sum_{j \in V''} w_j |\eta_j - t\zeta_j| = \sum_{j \in V''} w_j |\zeta_j| \left| t - \frac{\eta_j}{\zeta_j} \right|.$$

Thus $F(\gamma(t))$ is a piecewise linear function of t; it takes the minimum value at some $\bar{t} = \eta_\ell / \zeta_\ell$. From the assumption $\bar{t} \neq 0$ and from $\eta_\ell = \bar{t}\zeta_\ell$, we have $y_\ell - \langle \beta, z_\ell \rangle = \bar{t}\langle \rho, z_\ell \rangle$. That is,

$$y_\ell - \langle \beta + \bar{t}\rho, z_\ell \rangle = 0.$$

This means that we can decrease the value of the objective function by using $\mathsf{X}(V \cup \{\ell\})$ instead of $\mathsf{X}(V)$. Thus the lemma is proved. □

It now is clear that Lemma 4.4.1 is valid. We now proceed to prove Proposition 4.4.1.

The optimal solution for B is obtained by minimizing $J^i(\beta_i)$, $i = 1, \ldots, c$. If we put $w_k = (u_{ik})^m$, then $J^i(\beta_i)$ is equivalent to $F(\beta_i)$. From the above lemma the optimal solution β_i is obtained by solving the matrix equation

$$\mathsf{Y}(Z) = \mathsf{X}(Z)\beta$$

which has $p+1$ rows selected from $\{1, 2, \ldots, N\}$.

Let the subset of index of $p+1$ selected rows be $K = \{j_1, \ldots, j_{p+1}\} \subset \{1, 2, \ldots, N\}$ and the family of all such index sets be

$$\mathcal{K} = \{K : K = \{j_1, \ldots, j_{p+1}\} \subset \{1, 2, \ldots, N\}\}.$$

For an optimal solution B, c index sets $K_1, \ldots, K_c \in \mathcal{K}$ are thus determined. Now, suppose we have obtained the index sets $K_1, \ldots, K_c$; the same c sets $K_1, \ldots, K_c$ will not appear in the iteration of **FCRM** so long as the value of the objective function is strictly decreasing. In other words, if the same sets appears twice, the value of the objective function remains the same. Since all such combinations of subsets $K_1, \ldots, K_c$ are finite, the algorithm terminates after finite iterations of the major loop of **FCRM**.

4.4.2 An Illustrative Example

Let us observe a simple illustrative example that shows a characteristic of the method based on the least absolute deviation. Figure 4.6 depicts an artificial data set in which two regression models should be identified. Notice that there are five objects of outliers that may affect the models.

Figure 4.7 shows the result by the standard method of fuzzy c-regression models with $c = 2$ and $m = 2$. We observe the upper regression line is affected by the five outliers.

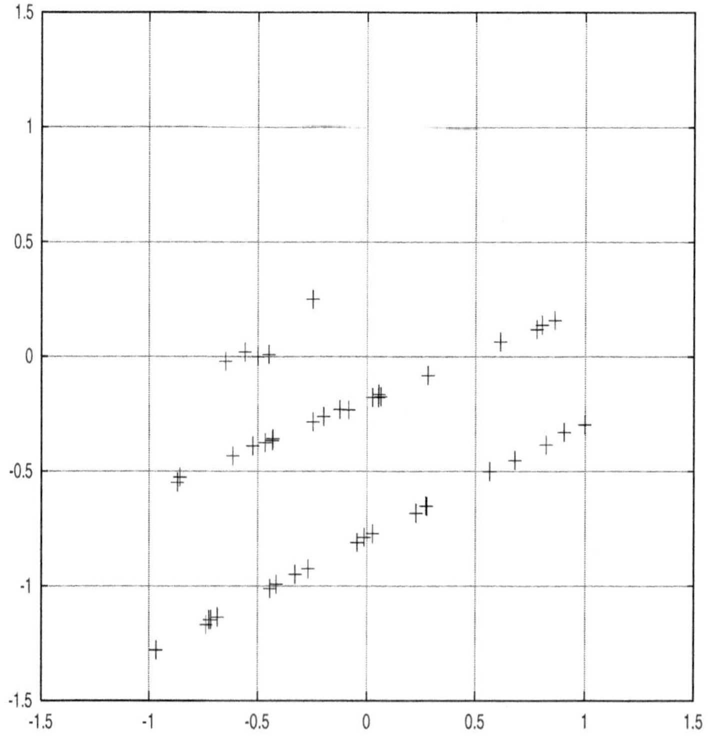

Fig. 4.6. An artificial data with outliers

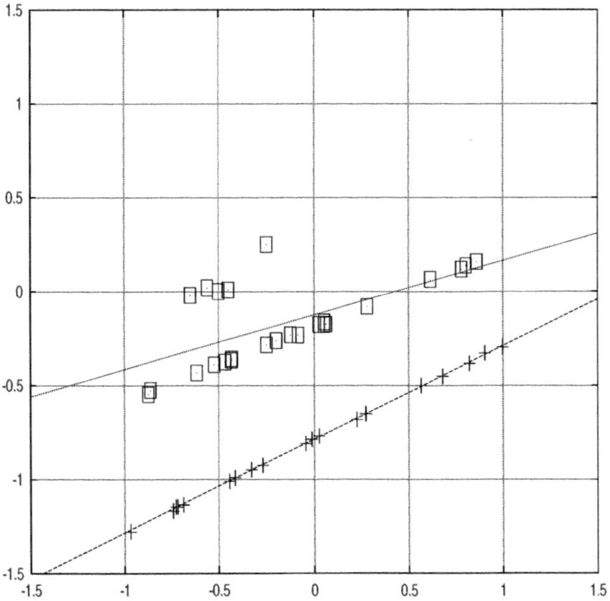

Fig. 4.7. Result by the standard method of fuzzy c-regression model based on the least square $(c = 2, m = 2)$

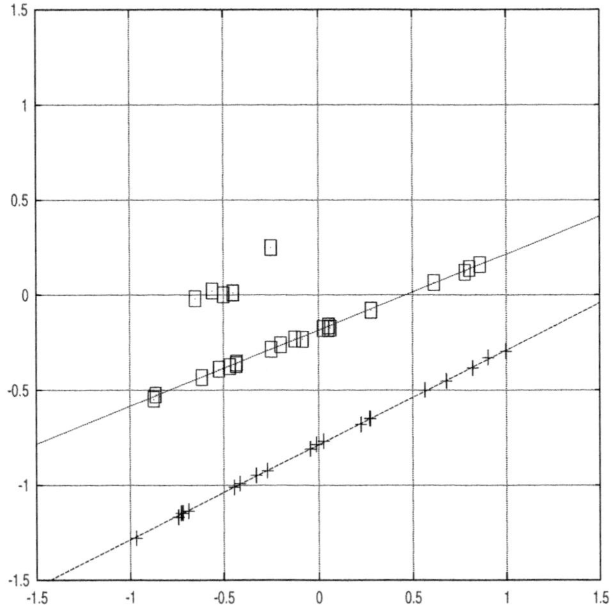

Fig. 4.8. Result by the standard method of fuzzy c-regression model based on the least absolute deviation $(c = 2, m = 2)$

Figure 4.8 shows the result by the standard method based on the least absolute deviation with $c = 2$ and $m = 2$. The upper regression line is not affected by the outliers.

The results by the entropy-based methods which are omitted here are the same as those by the standard methods.

5 Miscellanea

In this chapter we briefly note various studies on and around fuzzy c-means clustering that are not discussed elsewhere in this book.

5.1 More on Similarity and Dissimilarity Measures

In section 4.2, we have discussed the use of a similarity measure in fuzzy c-means. In the next section we mention some other methods of fuzzy clustering in which the Euclidean distance or other specific definitions of a dissimilarity measure is unnecessary. Rather, a measure $D(x_k, x_\ell)$ can be arbitrary so long as it has the interpretation of the dissimilarity between two objects. Moreover, a function $D(x, y)$ of variables (x, y) is unnecessary and only its value $D_{k\ell} = D(x_k, x_\ell)$ on an arbitrary pair of objects in X is needed. In other words, the algorithm works on the $N \times N$ matrix $[D_{k\ell}]$ instead of a binary relation $D(x, y)$ on $\boldsymbol{R}^p$.

Let us turn to the definition of dissimilarity measures. Suppose we are given a metric $D_0(x, y)$ that satisfies the three axioms including the triangular inequality. We do not care what type the metric $D_0(x, y)$ is, but we define

$$D_1(x, y) = \frac{D_0(x, y)}{1 + D_0(x, y)}.$$

We easily find that $D_1(x, y)$ is also a metric. To show this, notice that

(i) $D_1(x, y) \geq 0$ and $D_1(x, y) = 0 \iff x = y$ from the corresponding property of $D_0(x, y)$.
(ii) It is obvious to see $D_1(x, y) = D_1(y, x)$ from the symmetry of $D_0(x, y)$.
(iii) We observe the triangular inequality $D_1(x, y) + D_1(y, z) - D_1(x, z) \geq 0$ holds from straightforward calculation using $D_0(x, y) + D_0(y, z) - D_0(x, z) \geq 0$.

Note also that

$$0 \leq D(x, y) \leq 1, \qquad \forall x, y$$

This shows that given a metric, we can construct another metric that is bounded into the unit interval.

S. Miyamoto et al.: Algorithms for Fuzzy Clustering, STUDFUZZ 229, pp. 99–117, 2008.
springerlink.com

Another nontrivial metric which is a generalization of the Jaccard coefficient [1] is given as follows [99]:

$$D_J(x,y) = \frac{\sum_{j=1}^{p} |x^j - y^j|}{\sum_{j=1}^{p} \max\{x^j, y^j\}}, \qquad \forall x \geq 0,\ y \geq 0.$$

The proof of the triangular inequality is given in [99] and is omitted here.

Furthermore, we note that the triangular inequality is unnecessary for a dissimilarity measure in clustering. We thus observe that there are many more possible measures of dissimilarity for clustering.

Apart from the squared Euclidean distance and the L_1 distance, the calculation of cluster centers that minimize an objective function of the fuzzy c-means types using such a general dissimilarity is difficult. In the latter case we have two approaches:

1. Use a K-medoid clustering [80].
2. Use a relational clustering including agglomerative hierarchical clustering [1, 35, 99].

We do not discuss the K-medoid clustering here, as we cannot yet say definitely how and why this method is useful; the second approach of relational clustering is mentioned below.

5.2 Other Methods of Fuzzy Clustering

Although most researches on fuzzy clustering are concentrated on fuzzy c-means and its variations, there are still other methods using the concept of fuzziness in clustering.

The fuzzy equivalence relation obtained by the transitive closure of fuzzy symmetric and reflexive relations is one of these [99, 106], but it will be mentioned in the next section in its relation to agglomerative hierarchical clustering, since the method is very different from fuzzy c-means in nature.

In this section we show a few other methods that are more or less similar or related to fuzzy c-means.

5.2.1 Ruspini's Method

Ruspini's method [139, 140, 141] is one of the earliest algorithms for fuzzy clustering. We will briefly describe one of his algorithms [140]. Assume $X = \{x_1, \ldots, x_N\}$ is the set of objects as usual, and $D(x,y)$ is a dissimilarity measure. Unlike fuzzy c-means, it is unnecessary to restrict $D(x,y)$ to the squared Euclidean distance, nor to assume another form of a metric in the method of Ruspini; instead, we are simply given the matrix $[D(x_i, x_j)]$, $1 \leq i, j \leq N$. This method uses the membership matrix (u_{ki}) similar to that in fuzzy c-means in order to minimize an objective function, which is a second feature similar to fuzzy c-means.

The objective function is very different from that in fuzzy c-means:

$$J_{\text{Ruspini}}(U,\delta) = \sum_{i=1}^{c} \sum_{1\le j,k\le n} \{\delta(u_{ji} - u_{ki})^2 - D(x_j, x_k)\}^2 \tag{5.1}$$

in which the variables are $U = (u_{ki})$ and δ; the latter is a scalar variable. Thus the objective function is to approximate the dissimilarity $D(x_j, x_k)$ by the difference $(u_{ji} - u_{ki})^2$ and a scale variable δ. The minimization of $J_{\text{Ruspini}}(U, \delta)$ subject to $U \in \mathcal{U}_f$ is considered.

Iterative solutions using the gradient technique is described [140] but we omit the detail. Two important features in Ruspini's method is that an arbitrary dissimilarity measure can be handled and the method does not use a cluster center. These features are inherited to algorithms of *relational clustering* below.

5.2.2 Relational Clustering

The third chapter of [8] discusses relational clustering by agglomerative hierarchical algorithm which we will describe later, and by objective function algorithms. Another well-known method of relational clustering is FANNY [80] implemented in S-PLUS. We briefly mention two methods using objective functions [137, 80].

For this purpose let $D_{jk} = D(x_j, x_k)$ be an arbitrary dissimilarity measure and $U = (u_{ki})$ is the membership of the fuzzy partition $\mathcal{U}_f$ as usual. We consider the next two objective functions to be minimized with respect to U:

$$J_{\text{Roubens}}(U) = \sum_{i=1}^{c} \sum_{k=1}^{N} \sum_{\ell=1}^{N} u_{ki}^2 u_{\ell i}^2 D_{k\ell}, \tag{5.2}$$

$$J_{\text{FANNY}}(U) = \sum_{i=1}^{c} \frac{\sum_{k=1}^{N} \sum_{\ell=1}^{N} u_{ki}^2 u_{\ell i}^2 D_{k\ell}}{\sum_{j=1}^{N} u_{ji}^2}. \tag{5.3}$$

Notice that the both objective functions do not use a cluster center.

Let us consider $J_{\text{Roubens}}(U)$, if we define

$$\hat{D}_{ki} = \sum_{\ell=1}^{N} u_{\ell i}^2 D_{k\ell}, \tag{5.4}$$

then we have

$$J_{\text{Roubens}}(U) = \sum_{i=1}^{c} u_{ki}^2 \hat{D}_{ki},$$

which appears the same as the objective function of the fuzzy c-means. Hence the next iterative algorithm is used with a randomly given initial value of U.

I. Calculate $\hat{D}_{ki}$ by (5.4).

II. Calculate

$$u_{ki} = \left(\sum_{j=1}^{c} \frac{\hat{D}_{kj}}{\hat{D}_{ki}} \right)^{-1}$$

and if convergent, stop; else go to step **I**.

The algorithm of FANNY using $J_{\text{FANNY}}(U)$ is rather complicated and we omit the detail (see [80]). It aims at strict minimization but the solution does not have an explicit form but requires iteration.

There are also other methods of relational clustering using objective functions. Windham [169] considers

$$J_{\text{Windham}}(U,T) = \sum_{i=1}^{c}\sum_{k=1}^{N}\sum_{\ell=1}^{N} u_{ki}^2 t_{\ell i}^2 D_{k\ell},$$

which is similar to and yet different from J_{Roubens} in the sense that u_{ki} and $t_{\ell i}$ are two different sets of variables. As a result his algorithm repeats calculation of u_{ki} and $t_{\ell i}$ until convergence:

$$u_{ki} = \frac{1/\sum_{\ell} t_{\ell i}^2 D_{k\ell}}{\sum_j 1/\sum_{\ell} t_{\ell j}^2 D_{k\ell}},$$

$$t_{\ell i} = \frac{1/\sum_{k} u_{ki}^2 D_{k\ell}}{\sum_j 1/\sum_{k} u_{kj}^2 D_{k\ell}}.$$

Hathaway *et al.* [50] propose the relational fuzzy c-means using the next objective function:

$$J_{\text{Hathaway}}(U) = \frac{\sum_{k=1}^{N}\sum_{\ell=1}^{N} u_{ki}^m u_{\ell i}^m D_{k\ell}}{\sum_{j=1}^{N} u_{ji}^m},$$

which is a generalization of that in FANNY. His iteration algorithm is different from FANNY in nature and similar to that of fuzzy c-means. We omit the detail (see [50] or Chapter 3 of [8]).

5.3 Agglomerative Hierarchical Clustering

As noted in the introductory part of this book, the method of agglomerative hierarchical clustering which is very old has been employed in a variety of fields due to its usefulness. To describe agglomerative hierarchical clustering in detail is not the main purpose of his book, and hence we briefly review a part of the theory of agglomerative clustering in this section.

We first review the terms in agglomerative clustering for this purpose. As is usual, $X = \{x_1, \dots, x_N\}$ is the set of objects for clustering. In agglomerative clustering, however, X is not necessarily in a vector space and simply a set of objects. A measure of dissimilarity $D(x.x')$ is defined on an arbitrary pair $x, x' \in X$. As we noted above, no particular assumption on the dissimilarity is necessary for a number of methods in agglomerative clustering. In contrast, for a few other methods, the Euclidean space is assumed as we mention later.

$G_1, \dots, G_K$ are clusters that form the partition of X:

$$\bigcup_{i=1}^{K} G_i = X, \qquad G_i \cap G_j = \emptyset \quad (i \neq j),$$

as in the case of the crisp c-means. For a technical reason, we define a set of clusters by $\mathcal{G} = \{G_1, \dots, G_K\}$.

While K is a given constant in the crisp c-means, K varies as the algorithm of an agglomerative clustering proceeds. That is, the algorithm starts from the initial state where each object forms a cluster: $K = N$. In each iteration of the main loop of the algorithm, two clusters of the minimum dissimilarity are merged and the number of clusters is reduced by 1: $K = K - 1$, and finally all clusters are merged into the trivial cluster of X: $K = 1$.

In order to merge two clusters, an inter-cluster dissimilarity measure $D(G_i, G_j)$ is necessary of which different definitions are used. We next give a formal procedure of the agglomerative clustering in general and after that, different measures of inter-cluster dissimilarity are discussed.

Algorithm AHC: *Agglomerative Hierarchical Clustering.*

AHC1. Assume that initial clusters are given by
$\mathcal{G} = \{G_1, G_2, \dots, G_N\}$, $G_j = \{x_j\} \subset X$.
Set $K = N$ (K is the number of clusters).
Calculate $D(G, G')$ for all pairs $G, G' \in \mathcal{G}$ by $D(G, G') = D(x, x')$.

AHC2. Search the pair of minimum dissimilarity:

$$(G_p, G_q) = \arg \min_{G, G' \in \mathcal{G}} D(G, G'). \tag{5.5}$$

and let

$$m_K = D(G_p, G_q) = \min_{G, G' \in \mathcal{G}} D(G, G'). \tag{5.6}$$

Merge: $G_r = G_p \cup G_q$.
Add G_r to $\mathcal{G}$ and delete G_p, G_q from $\mathcal{G}$.
$K = K - 1$.
if $K = 1$ then stop.

AHC3. Update dissimilarity $D(G_r, G'')$ for all $G'' \in \mathcal{G}$.
Go to AHC2.

End AHC.

In **AHC**, the detail of constructing a dendrogram is omitted (see e.g., [99, 105]).

Different definitions for updating inter-cluster dissimilarity $D(G_r, G'')$ in AHC3 lead to different methods of agglomerative clustering: the single link, the complete link, and the average link are well-known methods usable for general dissimilarity measures.

- the single link (SL)

$$D(G, G') = \min_{x \in G, x' \in G'} D(x, x')$$

- the complete link (CL)

$$D(G, G') = \max_{x \in G, x' \in G'} D(x, x')$$

- the average link (AL)

$$D(G, G') = \frac{1}{|G||G'|} \sum_{x \in G, x' \in G'} D(x, x')$$

An important issue in agglomerative clustering is efficient updating of a dissimilarity measure. Ordinary methods such as the single link, complete link, average link, *etc.* have respective updating formulas using the dissimilarity matrix $[D_{ij}]$ in which $D_{ij} = D(G_i, G_j)$ and after merging, $D_{rj} = D(G_r, G_j)$ can be calculated solely from $D(G_p, G_j)$ and $D(G_q, G_j)$ instead of the above basic definitions. Namely, the single link and the complete link respectively use

$$D(G_r, G'') = \min\{D(G_p, G''), D(G_q, G'')\} \tag{5.7}$$

and

$$D(G_r, G'') = \max\{D(G_p, G''), D(G_q, G'')\} \tag{5.8}$$

and the average link uses

$$D(G_r, G'') = \frac{|G_p|}{|G_r|} D(G_p, G'') + \frac{|G_q|}{|G_r|} D(G_q, G'') \tag{5.9}$$

where $|G_r| = |G_p| + |G_q|$.

While the above three methods do not assume a particular type of a dissimilarity measure, there are other methods using the Euclidean space, one of which is the centroid method. Let us denote the centroid of a cluster by

$$v(G) = \frac{1}{|G|} \sum_{x \in G} x$$

in this section. The centroid method uses the squared Euclidean distance between the two centroids:

$$D(G, G') = \|v(G) - v(G')\|^2.$$

It is unnecessary to employ this definition directly to update $D(G_r, G'')$ in AHC3, since we have

$$D(G_r, G'') = \frac{|G_p|}{|G_r|} D(G_p, G'') + \frac{|G_q|}{|G_r|} D(G_q, G'') - \frac{|G_p||G_q|}{|G_r|^2} D(G_p, G_q) \tag{5.10}$$

for the centroid method [99, 105].

There is also the Ward method based on the Euclidean space which is said to be useful in many applications. The Ward method first defines the squared sum of errors around the centroid within a cluster:

$$E(G) = \sum_{x \in G} \|x - v(G)\|^2.$$

When two clusters are merged, the errors will increase, and the measure of the increase of the squared sum of errors is denoted by

$$\Delta(G, G') = E(G \cup G') - E(G) - E(G').$$

We have $\Delta(G, G') \geq 0$ for all $G, G' \in \mathcal{G}$. The Ward method uses this measure of increase as the dissimilarity:

$$D(G, G') = \Delta(G, G').$$

Notice that the initial value of $D(G, G')$ for $G = \{x\}, G' = \{x'\}$ is

$$D(G, G') = \frac{1}{2}\|x - x'\|^2$$

from the definition. Thus the Ward method chooses the pair that has the minimum increase of the sum of errors within the merged cluster. The updating formula to calculate $D(G_r, G'')$ solely from $D(G_p, G'')$, $D(G_q, G'')$, and $G(G_p, G_q)$ can also be derived. We have

$$\begin{aligned} D(G_r, G'') = & \frac{1}{|G_r| + |G''|}\{(|G_p| + |G''|)D(G_p, G'') + (|G_q| + |G''|)D(G_q, G'') \\ & - |G''|D(G_p, G_q)\}. \end{aligned} \tag{5.11}$$

The derivation is complicated and we omit the detail [1, 99, 105].

Note 5.3.1. For the single link, complete link, and average link, a similarity measure $S(x, x')$ can be used instead of a dissimilarity. In the case of a similarity, equation (5.5) should be replaced by

$$(G_p, G_q) = \arg \max_{G, G' \in \mathcal{G}} S(G, G'); \tag{5.12}$$

the rest of **AHC** is changed accordingly. It should also be noted that the definitions of the single link, the complete link, and the average link should be

- the single link (SL): $S(G, G') = \max_{x \in G, x' \in G'} S(x, x')$,
- the complete link (CL): $S(G, G') = \min_{x \in G, x' \in G'} S(x, x')$,
- the average link (AL): $S(G, G') = \dfrac{1}{|G||G'|} \displaystyle\sum_{x \in G, x' \in G'} S(x, x')$,

where we have no change in the average link. For the formula of updating in AHC3,

- the single link: $S(G_r, G'') = \max\{S(G_p, G''), S(G_q, G'')\}$,
- the complete link: $S(G_r, G'') = \min\{S(G_p, G''), S(G_q, G'')\}$,

as expected, while we have no change in the updating formula of the average link. For the centroid method and the Ward method, the use of a similarity measure is, of course, impossible.

Note 5.3.2. Kernel functions can also be used for agglomerative hierarchical clustering whereby a kernelized centroid method and a kernelized Ward method are defined [34]. An interesting fact is that the updating formulas when a kernel is used are just the same as the ordinary formula without a kernel, i.e., they are given by (5.10) and (5.11), respectively.

5.3.1 The Transitive Closure of a Fuzzy Relation and the Single Link

It is well-known that the transitive closure of a fuzzy reflexive and symmetric relation generates a hierarchical classification, in other words, a fuzzy equivalence relation [176]. We briefly note an important equivalence between the transitive closure and the single link.

Let us use a similarity measure $S(x_i, x_j)$, $x_i, x_j \in X$, and assume that

$$S(x,x) = \max_{x' \in X} S(x,x') = 1,$$
$$S(x,x') = S(x',x)$$

as usual. This assumption means that $S(x,x')$ as a relation on $X \times X$ is reflexive and symmetric.

We review the max-min composition of two fuzzy relations: suppose R_1 and R_2 are two fuzzy relations on $X \times X$. The max-min composition $R_1 \circ R_2$ is defined by

$$(R_1 \circ R_2)(x,z) = \max_{y \in X} \min\{R_1(x,y), R_2(y,z)\},$$

or, if we use the infix notation,

$$(R_1 \circ R_2)(x,z) = \bigvee_{y \in X} \{R_1(x,y) \wedge R_2(y,z)\}.$$

We write $S^2 = S \circ S$ and $S^n = S^{n-1} \circ S$ for simplicity.

Now, the transitive closure of S denoted by $\bar{S}$ is defined by

$$\bar{S} = S \vee S^2 \vee \cdots S^n \vee \cdots$$

When we have N elements in X, in other words, S is an $N \times N$ matrix, we have

$$\bar{S} = S \vee S^2 \vee \cdots S^N$$

It has been well-known that the transitive closure is reflexive, symmetric, and transitive:

$$\bar{S}(x,z) \geq \bigvee_{y \in X} \{\bar{S}(x,y) \wedge \bar{S}(y,z)\},$$

in other words, $\bar{S}$ is a fuzzy equivalence relation.

Let us consider an arbitrary α-cut $[R]_\alpha$ of a fuzzy equivalence relation R:

$$[R]_\alpha(x,y) = \begin{cases} 1 & (R(x,y) \geq \alpha), \\ 0 & (R(x,y) < \alpha). \end{cases}$$

It is also well-known that $[\bar{S}]_\alpha$ provides a partition by defining clusters $G_1(\alpha)$, $G_2(\alpha), \ldots$:

- x and y are in the same cluster $G_i(\alpha)$ if and only if $[\bar{S}]_\alpha(x,y) = 1$;
- x and y are in different clusters, i.e., $x \in G_i(\alpha)$ and $y \in G_j(\alpha)$ if and only if $[\bar{S}]_\alpha(x,y) = 0$.

It is moreover clear that if α increases, the partition becomes finer, since $[\bar{S}]_\alpha (x,y) = 1$ is less likely; conversely, when α decreases, the partition is coarser, since $[\bar{S}]_\alpha(x,y) = 1$ becomes easier. Thus we have a hierarchical classification by moving α in the unit interval. The hierarchical classification is formally given as follows. Let us define the set of clusters by

$$\mathcal{C}(\alpha) = \{G_1(\alpha), G_2(\alpha), \ldots\}$$

Take $\alpha \leq \alpha'$, we can prove

$$\forall G(\alpha) \in \mathcal{C}(\alpha),\ \exists G'(\alpha') \in \mathcal{C}(\alpha') \text{ such that } G'(\alpha') \subseteq G(\alpha),$$

although the proof is omitted to save space.

It has been shown that the hierarchical classification is equivalent to the single link applied to the same similarity measure $S(x,x')$. To establish the equivalence, we define the clusters obtained at the level α by the single link. Let us note the level where G_p and G_q are merged is given by m_K. When a similarity measure and the single link are used, we have the monotonic property:

$$m_N \geq m_{N-1} \geq \cdots \geq m_2.$$

The clusters at α in AHC are those formed at m_K that satisfy

$$m_K \geq \alpha > m_{K-1}.$$

We temporarily write the set of clusters formed at α by the single link method in AHC by $\mathcal{G}_{SL}(\alpha)$.

We now have the following, although we omit the proof.

Proposition 5.3.1. *For an arbitrary $\alpha \in [0,1]$, the set of clusters formed by the transitive closure and the single link are equivalent:*

$$\mathcal{C}(\alpha) = \mathcal{G}_{SL}(\alpha).$$

Note that the transitive closure is algebraically defined while the single link is based on algorithm. We thus obtain equivalence between the results by the algebraic and algorithmic methods that is far from trivial.

Note 5.3.3. A more comprehensive theorem of the equivalence among four methods including the transitive closure and the single link is stated in [99] where the rigorous proof is given. The key concept is the connected components of a fuzzy graph, although we omit the detail here.

5.4 A Recent Study on Cluster Validity Functions

To determine whether or not obtained clusters are of good quality is a fundamental and the most difficult problem in clustering. This problem is called validation of clusters, but fundamentally we have no sound criterion to judge the quality of clusters, and therefore the problem is essentially ill-posed. Nevertheless, there are many studies and proposals in this issue due to the importance of the problem.

Before stating cluster validity measures, we note there are three different approaches to validating clusters, that is,

1. hypothesis testing on clusters,
2. application of a model selection technique, and
3. the use of cluster validity measures.

The first two are based on the framework of mathematical statistics, and the assumption of the model of probability distributions is used. Hence the cluster quality is judged according to the assumed distribution. While the first method of hypothesis testing is based on a single model, the model selection technique assumes a family of probabilistic models from which the best should be selected [30].

Since the present consideration is mostly on fuzzy models, the first two techniques are useless, although there are possibilities for further study as to combine fuzzy and probabilistic models whereby the first two approaches will be useful even in fuzzy clustering.

Let us turn to the third approach of cluster validity functions. There are many measures of cluster validity [6, 8, 63, 28] and many discussions. We do not intend to overview them, but the main purpose of this section is to consider two measures using kernel functions [163, 164]. For this purpose we first discuss two different types of cluster validity measures, and then a number of non-kernelized measures using distances.

In addition, we note that there are many purposes to employ cluster validity measures [8, 63, 28]; one of the most important applications is to estimate the number of clusters. Hence we show examples to estimate the number of clusters, and then other use of the validity functions.

5.4.1 Two Types of Cluster Validity Measures

Although we do not overview many validity functions, we observe two types of cluster validity measures. First type consists of those measures using U alone without the dissimilarity, i.e., geometry of clusters, while others employ geometric information.

Three typical measures [8] of the first type are the partition coefficient:

$$F(U;c) = \frac{1}{N}\sum_{i=1}^{c}\sum_{k=1}^{N}(u_{ki})^2, \tag{5.13}$$

the degree of separation:

$$\rho(U;c) = 1 - \bigvee_{k=1}^{N} \left(\bigwedge_{i=1}^{c} u_{ki} \right), \tag{5.14}$$

and the partition entropy:

$$E(U;c) = -\frac{1}{N} \sum_{i=1}^{c} \sum_{k=1}^{N} u_{ki} \log u_{ki}. \tag{5.15}$$

We should judge clusters are better when one of these measures are larger. These three measures which employ U alone are, however, of limited use when compared with those of the second type due to the next reason.

Proposition 5.4.1. *For any positive integer c, the measures $F(U;c)$, $\rho(U;c)$, and $E(U;c)$ all take their maximum values unity when and only when the membership matrix is crisp, i.e., for all $k \in \{1, 2, \ldots, N\}$, there exists a unique $i \in \{1, \ldots, c\}$ such that $u_{ki} = 1$ and $u_{kj} = 0$ for all $j \neq i$.*

Thus these measures judge crisp clusters are best for any number of clusters c. Note also that the clusters approach crisp when $m \to 1$ in the method of Dunn and Bezdek, and when $\nu \to 0$ in the entropy-based method.

Although we do not say the first type is totally useless (see [8] where the way how to use these measures is discussed), we consider the second type of measures where dissimilarity $D(x_k, v_i)$ is used, some of which are as follows.

The determinants of covariance matrices: The sum of the determinants of fuzzy covariance matrices for all clusters has been studied by Gath and Geva [38]. Namely,

$$W_{det} = \sum_{i=1}^{c} \sqrt{\det F_i} \tag{5.16}$$

is used, where F_i is the fuzzy covariance matrix for cluster i:

$$F_i = \frac{\sum_{k=1}^{N} (u_{ki})^m (x_k - v_i)(x_k - v_i)^T}{\sum_{k=1}^{N} (u_{ki})^m}. \tag{5.17}$$

The traces of covariance matrices: Another measure associated with W_{det} is the sum of traces:

$$W_{tr} = \sum_{i=1}^{c} \mathrm{tr} F_i \tag{5.18}$$

in which F_i is given by (5.17).

Xie-Beni's index: Xie and Beni [171] propose the next index for cluster validity.

$$XB = \frac{\sum_{i=1}^{c}\sum_{k=1}^{N}(u_{ki})^m D(x_k, v_i)}{N \min_{1\le i,j\le c} \|v_i - v_j\|^2} . \tag{5.19}$$

When the purpose of a validity measure is to determine an appropriate number of clusters, the number of clusters which minimizes W_{det}, W_{tr}, or XB is taken.

5.4.2 Kernelized Measures of Cluster Validity

Why should a kernelized measure of cluster validity [71] be studied? The answer is clear if we admit the usefulness of a kernelized algorithm. It has been shown that a kernelized algorithm can produce clusters having nonlinear boundaries which ordinary c-means cannot have [42, 107, 108].

The above measures W_{det} and XB are based on the p-dimensional data space, while a kernelized algorithm uses the high-dimensional feature space H. W_{det} and XB which are based on the original data space are inapplicable to a kernelized algorithm, and therefore W_{det} and XB should be kernelized. It is straightforward to use the kernel-based variation of Xie-Beni's index, while W_{det} should be replaced by another but related measure using the trace of the fuzzy covariance matrix.

Let us show the reason why the determinant should not be used. It should be noted that H is an infinite dimensional space in which N objects form a subspace of dimension N [14]. When we handle a large number of data, we should handle the same large size N of the covariance matrix. This induces a number of difficulties in calculating the determinant.

1. A determinant of a matrix of size N requires $O(N^3)$ calculation, and the order grows rapidly when N is large.
2. It is known that the calculation of a determinant induces large numerical error when the size N is large.
3. It is also known that the calculation of a determinant is generally ill-conditioned when the size N is large.

5.4.3 Traces of Covariance Matrices

From the above reason we give up the determinant and use the trace instead. We hence consider

$$KW_{tr} = \sum_{i=1}^{c} \operatorname{tr} KF_i, \tag{5.20}$$

where $\operatorname{tr} KF_i$ is the trace of

$$KF_i = \frac{\sum_{k=1}^{N} (u_{ki})^m (\Phi(x_k) - W_i)(\Phi(x_k) - W_i)^T}{\sum_{k=1}^{N} (u_{ki})^m} . \tag{5.21}$$

A positive reason why the trace is used is that it can easily be calculated. Namely, we have

$$\mathrm{tr}\, KF_i = \frac{1}{S_i} \sum_{k=1}^{N} (u_{ki})^m \|\Phi(x_k) - W_i\|_H^2 = \frac{1}{S_i} \sum_{k=1}^{N} (u_{ki})^m D_H(\Phi(x_k), W_i), \quad (5.22)$$

Note that $D_H(\Phi(x_k), W_i) = \|\Phi(x_k) - W_i\|_H^2$ is given by

$$D_H(\Phi(x_k), W_i) = K_{kk} - \frac{2}{S_i} \sum_{j=1}^{N} (u_{ji})^m K_{jk} + \frac{1}{(S_i)^2} \sum_{j=1}^{N} \sum_{\ell=1}^{N} (u_{ji} u_{\ell i})^m K_{j\ell}, \quad (5.23)$$

where

$$S_i = \sum_{k=1}^{N} (u_{ki})^m, \quad (5.24)$$

$$K_{j\ell} = K(x_j, x_\ell), \quad (5.25)$$

as in the kernelized fuzzy c-means algorithm.

It should be noticed that KW_{tr} is similar to, and yet different from the objective function value by the standard method of Dunn and Bezdek.

5.4.4 Kernelized Xie-Beni Index

Xie-Beni's index can also be kernelized. We define

$$KXB = \frac{\sum_{i=1}^{c} \sum_{k=1}^{N} (u_{ki})^m \|\Phi(x_k) - W_i\|^2}{N \min_{i,j} \|W_i - W_j\|^2} = \frac{\sum_{i=1}^{c} \sum_{k=1}^{N} (u_{ki})^m D_H(\Phi(x_k), W_i)}{N \times \min_{i,j} D_H(W_i, W_j)}. \quad (5.26)$$

Notice that $D_H(\Phi(x_k), W_i)$ is given by (5.23), while

$$\begin{aligned} D_H(W_i, W_j) =& \langle W_i, W_i \rangle - 2 \langle W_i, W_j \rangle + \langle W_j, W_j \rangle \\ =& \frac{1}{S_i^2} \sum_{k=1}^{N} (u_{ki})^{2m} K_{kk} - \frac{2}{S_i S_j} \sum_{k=1}^{N} \sum_{h=1}^{N} (u_{ki} u_{hj})^m K_{kh} \\ &+ \frac{1}{S_j^2} \sum_{h=1}^{N} (u_{hj})^{2m} K_{hh}. \end{aligned} \quad (5.27)$$

5.4.5 Evaluation of Algorithms

As noted above, a main objective of the cluster validity measures is to determine an appropriate number of clusters. For example, W_{det} for different numbers of clusters are compared and the number c giving the minimum value of W_{det} should be selected.

Although the measures proposed here are used for this purpose, other applications of the measures are also useful. A typical example is comparative evaluation of different clustering algorithms. There are a number of different algorithms of c-means: the crisp c-means, the Dunn and Bezdek fuzzy c-means, the entropy-based fuzzy c-means, and also the kernelized versions of the three algorithms.

There are two types to comparatively evaluate different algorithms. One is to apply different algorithms to a set of well-known benchmark data with true classifications. However, an algorithm successful to an application does not imply usefulness of this method in all applications.

Thus, more consideration to compare different algorithms is necessary. A good criterion is robustness or stability of an algorithm which includes

(A) sensitivity or stability of outputs with respect to different initial values, and

(B) sensitivity or stability of outputs with respect to parameter variations.

Such stability can be measured using variations of a measure for clustering. For example, the variance of the objective functions with respect to different initial values appears adequate. However, an objective function is dependent on a used method and parameters. Hence a unified criterion that is independent from algorithms should be used. Thus, we consider the variance of a validity measure.

For example, let $\bar{W}_{det}$ and $V(W_{det})$ means the average and the variance of W_{det}, respectively, with respect to different initial values. We mainly consider $V(KW_{tr})$ and $V(KXB)$ to compare stability of different algorithms in the next section. To be specific, let us denote a set of initial values by $\mathcal{IV}$ and each initial value by $init \in \mathcal{IV}$. Moreover $KW_{tr}(init)$ and $KXB(init)$ be the two measures given the initial value $init$. Then,

$$\overline{KW}_{tr} = \frac{1}{|\mathcal{IV}|} \sum_{init \in \mathcal{IV}} KW_{tr}(init),$$

$$V(KW_{tr}) = \frac{1}{|\mathcal{IV}| - 1} \sum_{init \in \mathcal{IV}} \left(KW_{tr}(init) - \overline{KW}_{tr}\right)^2.$$

$\overline{KXB}$ and $V(KXB)$ are defined in the same way.

5.5 Numerical Examples

We first consider a set of illustrative examples as well as well-known real examples and investigate the 'optimal' number of clusters judged by different measures. The illustrative examples include typical data that can only be separated by kernelized methods. Secondly, robustness of algorithms is investigated. Throughout the numerical examples the Gaussian kernel (4.3) is employed.

5.5.1 The Number of Clusters

Figures 5.1∼5.4 are illustrative examples whereby measures of cluster validity W_{det}, W_{tr}, XB, KW_{tr}, and KXB are tested. In these figures, each number c

Table 5.1. The numbers of clusters whereby the minimum of the validity measures have been obtained. (The symbol '-' in the table means that the value is monotonically decreasing as the number of clusters increases.)

	Fig.5.1	Fig.5.2	Fig.5.3	Fig.5.4	Iris
	number of clusters				
W_{det}	4	4	5	6	2
XB	4	4	5	5	2
W_{tr}	4	4	-	6	3
KXB	4	4	5	5	2
KW_{tr}	4	4	5	5	2

Table 5.2. The numbers of clusters whereby the minimum of the kernelized measures have been obtained. (The symbol '-' in the table means that the value is monotonically decreasing as the number of clusters increases.)

	two circles(Fig.5.5)	three circles(Fig.5.6)
	number of clusters	
KXB	6	-
KW_{tr}	2	2

Table 5.3. Evaluation of different algorithms by KW_{tr} using *Iris* data. (Processor time is in sec and miscl. implies misclassifications.)

algorithm	processor time	miscl.	average	variance
HCM	2.09×10^{-3}	18.27	0.151	5.56×10^{-5}
sFCM	1.76×10^{-2}	17.00	0.184	2.85×10^{-10}
eFCM	4.27×10^{-3}	16.79	0.152	4.40×10^{-5}
K-HCM	2.65×10^{-1}	16.68	0.164	1.04×10^{-3}
K-sFCM	1.39	17.33	0.187	1.27×10^{-4}
K-eFCM	1.66	14.00	0.175	2.48×10^{-11}

Table 5.4. Evaluation of different algorithms by KW_{tr} using *BCW* data. (Processor time is in sec and miscl. implies misclassifications.)

algorithm	processor time	miscl.	average	variance
HCM	8.75×10^{-3}	26.09	1.42	4.33×10^{-7}
sFCM	4.59×10^{-2}	28.00	1.54	3.21×10^{-12}
eFCM	1.53×10^{-2}	27.00	1.45	8.56×10^{-13}
K-HCM	1.87	21.00	1.38	4.93×10^{-30}
K-sFCM	8.97	20.00	1.63	2.41×10^{-13}
K-eFCM	4.78	21.00	1.39	1.60×10^{-14}

of clusters has been assumed and then the entropy-based clustering has been carried out, and the values of measures have been calculated. For a kernelized measure, the kernelized method using the entropy-based objective function has

Table 5.5. Evaluation of different algorithms by *KXB* using *Iris* data. (Processor time is in sec and miscl. implies misclassifications.)

algorithm	processor time	miscl.	average	variance
HCM	2.09×10^{-3}	18.27	0.138	9.84×10^{-3}
sFCM	1.76×10^{-2}	17.00	0.147	1.60×10^{-8}
eFCM	4.27×10^{-3}	16.79	0.125	1.71×10^{-3}
K-HCM	2.65×10^{-1}	16.68	0.165	2.46×10^{-2}
K-sFCM	1.39	17.33	0.150	1.10×10^{-3}
K-eFCM	1.66	14.00	0.158	6.34×10^{-3}

Table 5.6. Evaluation of different algorithms by *KXB* using *BCW* data. (Processor time is in sec and miscl. implies misclassifications.)

algorithm	processor time	miscl.	average	variance
HCM	8.75×10^{-3}	26.09	0.108	2.80×10^{-9}
sFCM	4.59×10^{-2}	28.00	0.124	5.05×10^{-15}
eFCM	1.53×10^{-2}	27.00	0.109	4.23×10^{-15}
K-HCM	1.87	21.00	0.113	1.93×10^{-28}
K-sFCM	8.97	20.00	0.165	6.79×10^{-14}
K-eFCM	4.78	21.00	0.113	6.94×10^{-16}

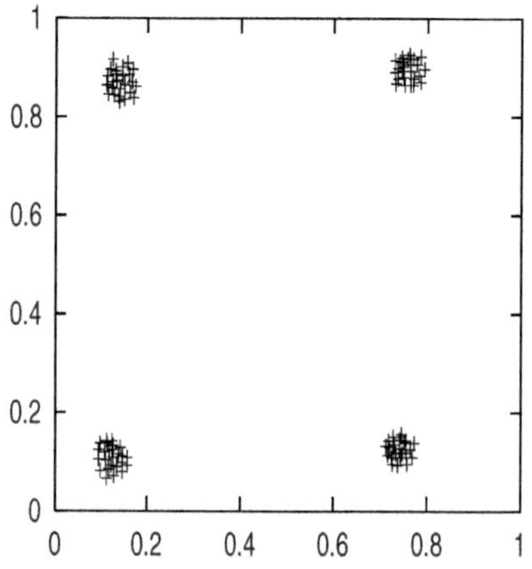

Fig. 5.1. An illustrative example of 4 clusters of 100 objects

been applied. Table 5.1 summarizes the number of clusters for which the minimum value of each measure has been obtained. Table 5.1 also includes outputs from the well-known *Iris* data. We note that all methods described here have no classification errors for the examples in Figures 5.1~5.4. For *Iris*, a linearly separated group from the other two has been recognized as a cluster when $c = 2$, and the number of misclassifications are listed in Table 5.3 when $c = 3$.

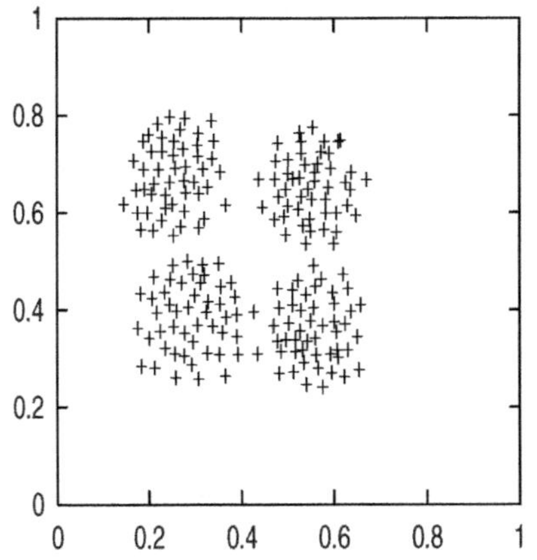

Fig. 5.2. An illustrative example of 4 clusters of 200 objects

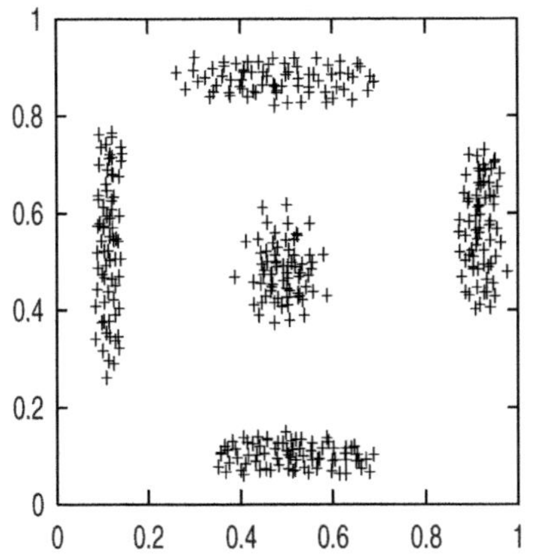

Fig. 5.3. An illustrative example of 5 clusters of 350 objects

From Table 5.1 we see that W_{tr} fails for the example in Fig.5.3, while other measures work well. We should remark that while W_{tr} is not better than other measures, KW_{tr} works well for all examples. It should also be noticed that most measures tell $c = 2$ is an appropriate number of clusters for *Iris* data [182]. It is known that a group in *Iris* data set is linearly separated from the other two, while the rest of the two groups are partly mixed. Hence whether $c = 2$ or $c = 3$ is an appropriate number of clusters is a delicate problem for which no one can have a definite answer.

Figures 5.5 and 5.6 are well-known examples for which ordinary algorithms do not work. For Fig.5.5, kernelized c-means algorithms as well as a kernelized LVQ clustering [70] can separate the outer circle and the inner ball [42, 107, 108]. For Fig. 5.6, it is still difficult to separate the two circles and the innermost

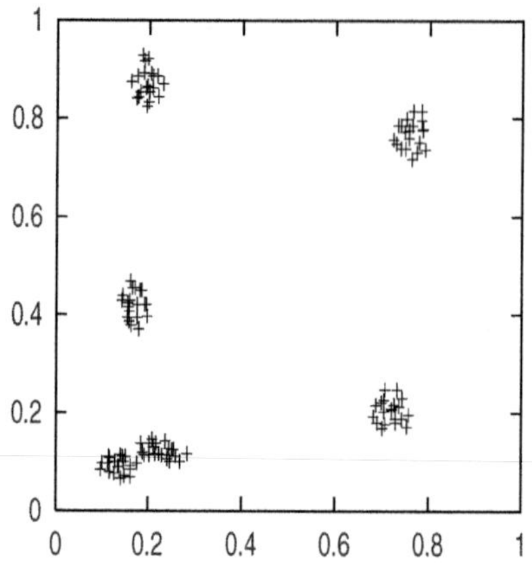

Fig. 5.4. An illustrative example of 5∼6 clusters of 120 objects

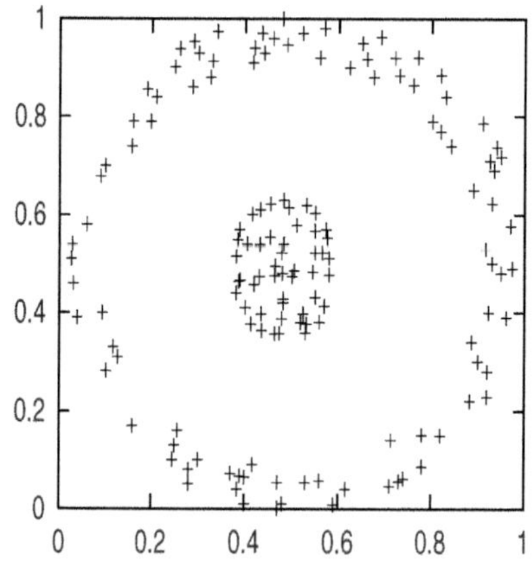

Fig. 5.5. The 'ring and ball' data

ball; only the kernelized LVQ clustering [70] has separated the three groups. We thus used the kernelized LVQ algorithm for different numbers of clusters for Figures 5.5 and 5.6. We have tested KW_{tr} and KXB. The other measures W_{det}, W_{tr}, XB have also been tested but their values are monotonically decreasing as the number of clusters increases.

Table 5.2 shows the number of clusters with the minimum values of KW_{tr} and KXB for the two examples. Other measures have been omitted from the above reason, in other words, they all will have '-' symbols in the corresponding cells. From the table we see that the two measure work correctly for Fig.5.5, while KW_{tr} judges $c = 2$ is appropriate and KXB does not work well for Fig.5.6.

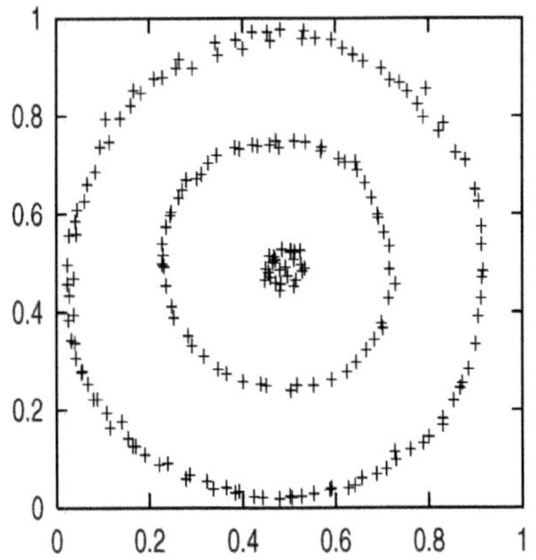

Fig. 5.6. The data of two rings and innermost ball

5.5.2 Robustness of Algorithms

Another application of these measures is comparison of robustness or stability of different algorithms.

Two well-known real data sets of *Iris* and *Wisconsin breast cancer* (abbreviated as *BCW*) [182] have been used for the test and the kernelized measures KW_{tr} and KXB have been employed. The compared algorithms are the basic crisp c-means (abbreviated HCM), the Dunn and Bezdek fuzzy c-means (abbreviated sFCM), the entropy-based fuzzy c-means (abbreviated eFCM), the kernelized crisp c-means (abbreviated K-HCM), the kernelized Dunn and Bezdek fuzzy c-means (abbreviated K-sFCM), and the kernelized entropy fuzzy c-means (abbreviated K-eFCM).

The measured quantities are:

- processor time in sec,
- the number of misclassifications,
- the average $\overline{KW_{tr}}$ or $\overline{KXB}$, and
- the variance $V(KW_{tr})$ or $V(KXB)$.

The average implies the quality of clusters judged by the measure, while the variance shows stability of an algorithm. The test has been carried out with 100 trials of different initial values.

Tables 5.3∼5.6 summarize the results. From the two tables for *Iris* we observe that sFCM is judged to be more stable than HCM and eFCM, while this tendency is not clear in *BCW*. Overall, the kernelized algorithms are judged to be more stable than ordinary algorithms.

Note 5.5.1. The kernelized Xie-Beni index has been presented by Inokuchi *et al.* [71] and by Gu and Hall [44] simultaneously at the same conference as independent studies.

6 Application to Classifier Design

This chapter is devoted to a description of the postsupervised classifier design using fuzzy clustering. We will first derive a modified fuzzy c-means clustering algorithm by slightly generalizing the objective function and introducing some simplifications. The k-harmonic means clustering [177, 178, 179, 119] is reviewed from the point of view of fuzzy c-means. In the algorithm derived from the iteratively reweighted least square technique (IRLS), membership functions are variously chosen and parameterized. Experiments on several well-known benchmark data sets show that the classifier using a newly defined membership function outperforms well-established methods, i.e., the support vector machine (SVM), the k-nearest neighbor classier (k-NN) and the learning vector quantization (LVQ). Also concerning storage requirements and classification speed, the classifier with modified **FCM** improves the performance and efficiency.

6.1 Unsupervised Clustering Phase

The clustering is used as an unsupervised phase of classifier designs. We first recapitulate the three kinds of objective functions, i.e., the standard, entropy-based, and quadratic-term-based fuzzy c-means clustering in Chapter 2. The objective function of the standard method is:

$$\bar{U} = \arg\min_{U \in \mathcal{U}_f} J_{\text{fcm}}(U, \bar{V}). \tag{6.1}$$

$$J_{\text{fcm}}(U, V) = \sum_{i=1}^{c} \sum_{k=1}^{N} (u_{ki})^m D(x_k, v_i), \quad (m > 1). \tag{6.2}$$

under the constraint:

$$\mathcal{U}_f = \{ U = (u_{ki}) : \sum_{j=1}^{c} u_{kj} = 1,\ 1 \le k \le N;$$
$$u_{ki} \in [0, 1],\ 1 \le k \le N,\ 1 \le i \le c \}. \tag{6.3}$$

$D(x_k, v_i)$ denotes the squared distance between x_k and v_i, so the standard objective function is the weighted sum of squared distances.

S. Miyamoto et al.: Algorithms for Fuzzy Clustering, STUDFUZZ 229, pp. 119–155, 2008.
springerlink.com

Following objective function is used for the entropy-based method.

$$J_{\text{efc}}(U,V) = \sum_{i=1}^{c}\sum_{k=1}^{N} u_{ki} D(x_k, v_i) + \nu \sum_{i=1}^{c}\sum_{k=1}^{N} u_{ki} \log u_{ki}, \quad (\nu > 0). \tag{6.4}$$

The objective function of the quadratic-term-based method is:

$$J_{\text{qfc}}(U,V) = \sum_{i=1}^{c}\sum_{k=1}^{N} u_{ki} D(x_k, v_i) + \frac{1}{2}\nu \sum_{i=1}^{c}\sum_{k=1}^{N} (u_{ki})^2, \quad (\nu > 0). \tag{6.5}$$

Combining these three, the objective function can be written as :

$$J(U,V) = \sum_{i=1}^{c}\sum_{k=1}^{N} (u_{ki})^m D(x_k, v_i) + \nu \sum_{i=1}^{c}\sum_{k=1}^{N} K(u), \tag{6.6}$$

where both m and ν are the fuzzifiers.

When $m > 1$ and $\nu = 0$, (6.6) is the standard objective function.

When m=1 and $K(u) = u_{ki}\text{log}u_{ki}$, (6.6) is the objective function of the entropy-based method. The algorithm is similar to the EM algorithm for the normal mixture or Gaussian mixture models whose covariance matrices are a unit matrix and cluster volumes are equal.

When $m = 1$ and $K(u) = \frac{1}{2}(u_{ki})^2$, (6.6) is the objective function of the quadratic-term-based method.

6.1.1 A Generalized Objective Function

From the above consideration, we can generalize the standard objective function. Let $m > 1$ and $K(u) = (u_{ki})^m$, then (6.6) is the objective function ($J_{\text{gfc}}(U,V)$) from which we can easily derive the necessary condition for the optimality.

We consider minimization of (6.6) with respect to U under the condition $\sum_{i=1}^{c} u_{ki} = 1$. Let the Langrange multiplier be λ_k, $k = 1, \ldots, N$, and put

$$\begin{aligned} L &= J_{\text{gfc}}(U,V) + \sum_{k=1}^{N} \lambda_k (\sum_{i=1}^{c} u_{ki} - 1) \\ &= \sum_{i=1}^{c}\sum_{k=1}^{N} (u_{ki})^m D(x_k, v_i) + \nu \sum_{i=1}^{c}\sum_{k=1}^{N} (u_{ki})^m \\ &+ \sum_{k=1}^{N} \lambda_k (\sum_{i=1}^{c} u_{ki} - 1). \end{aligned} \tag{6.7}$$

For the necessary condition of optimality of (6.7) we differentiate

$$\frac{\partial L}{\partial u_{ki}} = m(u_{ki})^{m-1}(D(x_k, v_i) + \nu) + \lambda_k = 0.$$

$D(x_k, v_i) + \nu > 0$ $(i = 1, \ldots, c)$, if $\nu > 0$. To eliminate λ_k, we note

$$u_{kj} = \left[\frac{-\lambda_k}{m(D(x_k, v_j) + \nu)} \right]^{\frac{1}{m-1}}. \tag{6.8}$$

Summing up for $j = 1, \ldots, c$ and taking $\sum_{j=1}^{c} u_{kj} = 1$ into account, we have

$$\sum_{j=1}^{c} \left[\frac{-\lambda_k}{m(D(x_k, v_j) + \nu)} \right]^{\frac{1}{m-1}} = 1.$$

Using this equation to (6.8), we can eliminate λ_k, having

$$u_{ki} = \left[\sum_{j=1}^{c} \left(\frac{D(x_k, v_i) + \nu}{D(x_k, v_j) + \nu} \right)^{\frac{1}{m-1}} \right]^{-1}. \tag{6.9}$$

This solution satisfies $u_{ki} \geq 0$ and u_{ki} is continuous if $\nu > 0$.

The optimal solution for V is also easily derived by differentiating L with respect to V.

$$v_i = \frac{\sum_{k=1}^{N} (u_{ki})^m x_k}{\sum_{k=1}^{N} (u_{ki})^m}. \tag{6.10}$$

We now have insight about the property, which distinguishes the generalized method from the standard and entropy-based methods. Let $\mathbf{U}_{\text{gfc}}^{(i)}(x; V)$ denote the classification function for the generalized method. We later use the term "discriminant function", which refers to a function for pattern classification. In this book, we use the term "classification function" to signify the function for partitioning the input feature space into clusters.

$$\mathbf{U}_{\text{gfc}}^{(i)}(x; V) = \left[\sum_{j=1}^{c} \left(\frac{D(x, v_i) + \nu}{D(x, v_j) + \nu} \right)^{\frac{1}{m-1}} \right]^{-1}. \tag{6.11}$$

It should be noted that $\mathbf{U}_{\text{gfc}}^{(i)}(x; V)$ has a close relationship with Cauchy weight function in M-estimation [53, 64] when $m = 2$ and $\nu = 1$. The membership function of the standard method suffers from the singularity which occurs when $D(x_k, v_i) = 0$. When $\nu > 0$, (6.9) alleviates the singularity. We will apply this type of weight function to a classifier design in the next section.

Next proposition states that ν is a fuzzifier.

Proposition 6.1.1. *The function $\mathbf{U}_{\text{gfc}}^{(i)}(x; V)$ is a decreasing function of ν when*

$$\|x - v_i\| < \|x - v_j\|, \ \forall j \neq i \quad (x \in \mathbf{R}^p),$$

and if $\nu > 0$,

$$\max_{x \in \mathbf{R}^p} \mathbf{U}_{\mathrm{gfc}}^{(i)}(x; V) = \mathbf{U}_{\mathrm{gfc}}^{(i)}(v_i; V) < 1. \tag{6.12}$$

$\mathbf{U}_{\mathrm{gfc}}^{(i)}(x; V)$ *is an increasing function of* ν *when*

$$\|x - v_i\| > \|x - v_j\|, \ \forall j \neq i \quad (x \in \mathbf{R}^p).$$

$\mathbf{U}_{\mathrm{gfc}}^{(i)}(x; V)$ *tends to* $1/c$ *as* $\nu \to +\infty$.

$$\lim_{\nu \to +\infty} \mathbf{U}_{\mathrm{gfc}}^{(i)}(x; V) = \frac{1}{c}. \tag{6.13}$$

$\mathbf{U}_{\mathrm{gfc}}^{(i)}(x; V)$ *tends to* $1/c$ *as* $\|x\| \to +\infty$.

$$\lim_{\|x\| \to +\infty} \mathbf{U}_{\mathrm{gfc}}^{(i)}(x; V) = \frac{1}{c}. \tag{6.14}$$

Proof. The first part of the proposition immediately follows from

$$1/\mathbf{U}_{\mathrm{gfc}}^{(i)}(x; V) - 1 = \sum_{j \neq i} \left(\frac{\|x - v_i\|^2 + \nu}{\|x - v_j\|^2 + \nu} \right)^{\frac{1}{m-1}}, \tag{6.15}$$

hence (6.12) follows.

For (6.13) and (6.14), we can use (6.15) again:

$$1/\mathbf{U}_{\mathrm{gfc}}^{(i)}(x; V) - 1 \to c - 1 \quad (\nu \to +\infty \text{ or } \|x\| \to +\infty).$$ □

$\mathbf{U}_{\mathrm{fcm}}^{(i)}(x; V)$ shares the last property (6.14) but

$$\max_{x \in \mathbf{R}^p} \mathbf{U}_{\mathrm{fcm}}^{(i)}(x; V) = \mathbf{U}_{\mathrm{fcm}}^{(i)}(v_i; V) = 1 \tag{6.16}$$

for $\forall i$, if all v_i's are distinctive. Table 6.1 summarizes the characteristics of the three methods when $m > 1$ and $\nu > 0$.

Deterministic annealing [135] is a method of clustering closely related to the entropy based **FCM** (i.e., **FCM**-e). During the iterations of the **FCM**-e algorithm, the phase transition occurs and the number of effective clusters increases by decreasing the value of fuzzifier ν. ν is referred to as "temperature" in the deterministic annealing. With the standard **FCM**, cluster centers tend to remain

Table 6.1. Characteristics of fuzzy clusterings

	$\|x\| \to +\infty$	$x \to v_i$
standard (FCM-s)	fuzzy	crisp
entropy-based (FCM-e)	crisp	crisp-fuzzy
generalized (FCM-g)	fuzzy	crisp-fuzzy

in place even when the fuzzifier ν is changed to a very large number. **FCM-s** is robust from this point of view, but does not cause the phase transition in the sense of deterministic annealing [135].

Note that the objective function of the possibilistic clustering [87, 23] is written similarly to (6.7) as:

$$J_{\text{pos}}(U,V) = \sum_{i=1}^{c}\sum_{k=1}^{N}(u_{ki})^m D(x_k, v_i) + \nu \sum_{i=1}^{c}\sum_{k=1}^{N}(1-u_{ki})^m \tag{6.17}$$

where the condition $\sum_{i=1}^{c} u_{ki} = 1$ is omitted. As pointed out in [23], the possibilistic clustering is closely related with robust M-estimation [53, 64] and ν in (6.17) plays the role of robustizer whereas ν in (6.7) is a fuzzifier as stated in Proposition 6.1.1.

6.1.2 Connections with k-Harmonic Means

The k-harmonic means (KHM) clustering [177, 178, 179, 119] is a relatively new iterative unsupervised learning algorithm. KHM is essentially insensitive to the initialization of the centers. It basically consists of penalizing solutions via weights on the data points, somehow making the centers move toward the hardest (difficult) points. The motivations come from an analogy with supervised classifier design methods known as boosting [84, 144, 82].

The harmonic average of c numbers $a_1, ..., a_c$ is defined as $\frac{c}{\sum_{i=1}^{c} \frac{1}{a_i}}$. For clarifying the connection between **FCM** and **KHM**, the objective function of **KHM** is rewritten as:

$$\begin{aligned} J_{\text{KHM}}(V) &= \sum_{k=1}^{N} \frac{c}{\sum_{i=1}^{c} \frac{1}{\|x_k - v_i\|^{\frac{2}{m-1}}}} \\ &= \sum_{k=1}^{N}\sum_{i=1}^{c} \left[\frac{D(x_k, v_i)^{\frac{1}{m-1}}}{\sum_{j=1}^{c} \left(\frac{D(x_k, v_i)}{D(x_k, v_j)} \right)^{\frac{1}{m-1}}} \right] \\ &= \sum_{k=1}^{N}\sum_{i=1}^{c} \left[\sum_{j=1}^{c} \left(\frac{D(x_k, v_i)}{D(x_k, v_j)} \right)^{\frac{1}{m-1}} \right]^{-1} D(x_k, v_i)^{\frac{1}{m-1}}. \end{aligned} \tag{6.18}$$

When m=2, (6.18) is the same as J_{fcm} in (6.2) with $m = 1$, where we substitute with (6.9), $m = 2$ and $\nu = 0$. m must be greater than 1 for the standard method, so the objective function does not coincide with the standard method, though as we will see below the update rule of centers v_i is the same as (6.10) with $m = 2$.

By taking partial derivative of $J_{\text{KHM}}(V)$ with respect to v_i, we have

$$\frac{\partial J_{\text{KHM}}(V)}{\partial v_i} = -cm \sum_{k=1}^{N} \frac{x_k - v_i}{D(x_k, v_i)^{\frac{1}{m-1}+1} \left(\sum_{j=1}^{c} \frac{1}{D(x_k, v_j)^{\frac{1}{m-1}}} \right)^2} = 0. \tag{6.19}$$

Although $D(x_k, v_i)$ includes v_i, from (6.19) the iterative update rule can be written as:

$$v_i = \frac{\sum_{k=1}^{N} \frac{1}{D(x_k, v_i)^{\frac{1}{m-1}+1} \left(\sum_{j=1}^{c} \frac{1}{D(x_k, v_j)^{\frac{1}{m-1}}} \right)^2} x_k}{\sum_{k=1}^{N} \frac{1}{D(x_k, v_i)^{\frac{1}{m-1}+1} \left(\sum_{j=1}^{c} \frac{1}{D(x_k, v_j)^{\frac{1}{m-1}}} \right)^2}}$$

$$= \frac{\sum_{k=1}^{N} \left[\sum_{j=1}^{c} \left(\frac{D(x_k, v_i)}{D(x_k, v_j)} \right)^{\frac{1}{m-1}} \right]^{-2} D(x_k, v_i)^{\frac{1}{m-1}-1} x_k}{\sum_{k=1}^{N} \left[\sum_{j=1}^{c} \left(\frac{D(x_k, v_i)}{D(x_k, v_j)} \right)^{\frac{1}{m-1}} \right]^{-2} D(x_k, v_j)^{\frac{1}{m-1}-1}} \tag{6.20}$$

When m=2, (6.20) is the same as (6.10) substituted with (6.9) where $m = 2$ and $\nu = 0$. Thus, we have the same clustering results with the standard method when $m = 2$. In (6.20), $\left[\sum_{l=1}^{c} \left(\frac{D(x_k, v_i)}{D(x_k, v_l)} \right)^{\frac{1}{m-1}} \right]^{-2}$ is the weight on x_k for computing weighted mean of x_k's.

Let u_{ki} be the membership function as:

$$u_{ki} = \left[\sum_{j=1}^{c} \left(\frac{D(x_k, v_i)}{D(x_k, v_j)} \right)^{\frac{1}{m-1}} \right]^{-1}, \tag{6.21}$$

then for $\forall m > 0$, u_{ki}'s sum to one except when $D(x_k, v_i) = 0$.

$$\sum_{i=1}^{c} u_{ki} = \sum_{i=1}^{c} \left[\sum_{j=1}^{c} \left(\frac{D(x_k, v_i)}{D(x_k, v_j)} \right)^{\frac{1}{m-1}} \right]^{-1} = 1 \tag{6.22}$$

$D(x_k, v_i)^{\frac{1}{m-1}-1}$ in (6.20) can also be seen as a weight on a data point, which comes from an analogy with supervised classifier design methods known as boosting. As $D(x_k, v_i)$ approaches to zero, the effect of u_{ki} for computing v_i decreases when $m < 2$. This view of the weights is slightly different from [177, 178, 179] but the effect of the weights is the same. Similar to the standard method, KHM clustering also suffers from the singularity which occur when $D(x_k, v_i) = 0$. The weight $D(x_k, v_i)^{\frac{1}{m-1}-1}$ mitigates its effect when $m < 2$.

6.1.3 Graphical Comparisons

Characteristics of the four clustering methods are compared in Figs.6.1-6.8, where $c = 3$ and other parameter values are given in the legend of each figure. In the figures, upper left and middle, and lower left graphs show 3D graphics of the classification functions. Lower middle graph shows the function (u^*) of distance from a cluster center.

$$(\text{FCM}-\text{s}) \qquad u^*_{ki} = \frac{1}{(D(x_k, v_i))^{\frac{1}{m-1}}}, \tag{6.23}$$

$$(\text{FCM}-\text{e}) \qquad u^*_{ki} = \exp\left(-\frac{D(x_k, v_i)}{\nu}\right), \tag{6.24}$$

$$(\text{FCM}-\text{g}) \qquad u^*_{ki} = \frac{1}{(D(x_k, v_i) + \nu)^{\frac{1}{m-1}}}. \tag{6.25}$$

$$(\text{KHM}) \qquad u^*_{ki} = \frac{1}{(D(x_k, v_i))^{\frac{1}{m-1}}}. \tag{6.26}$$

Upper right graph shows the contours of classification functions. The contours of maximum values between the three classification functions are drawn. Lower right graph shows the clustering results where stars mark cluster centers.

Figs.6.1 and 6.2 show rather crisply partitioned results by the standard method (**FCM**-s) with $m = 1.3$ and by the entropy-based method (**FCM**-e) with $\nu = 0.05$ respectively. The contours are different from each other at the upper right corner of the graphs since the points near (1.0, 1.0) are far from all the cluster centers. Figs.6.3 and 6.4 show fuzzily partitioned results by **FCM**-s with $m = 4$ and by **FCM**-e with $\nu = 0.12$ respectively. These two methods produce quite different contours of classification functions when the fuzzifier is relatively large. Fig.6.3 shows the robustness of **FCM**-s where all centers are located at densely accumulated areas. **FCM**-s suffers from the problem called singularity when $D(x_k, v_i) = 0$, which thus results in a singular shape of the classification function $\mathbf{U}^{(i)}_{\text{fcm}}(x; V)$. When the fuzzifier m is large, the classification function appears to be spiky at the centers as shown by 3D graphics in Fig.6.3. This is the singularity in shape.

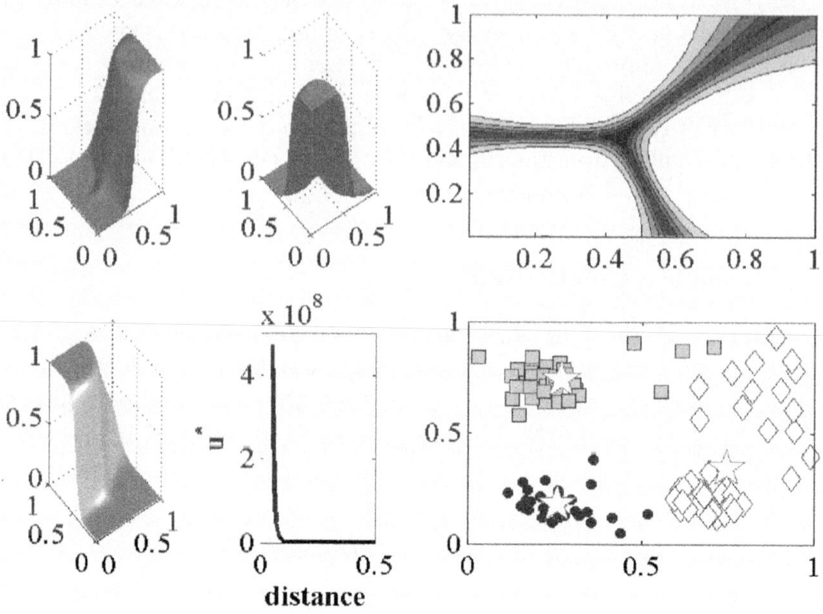

Fig. 6.1. Rather crisply partitioned result by the standard method (**FCM**-s) with $m = 1.3$

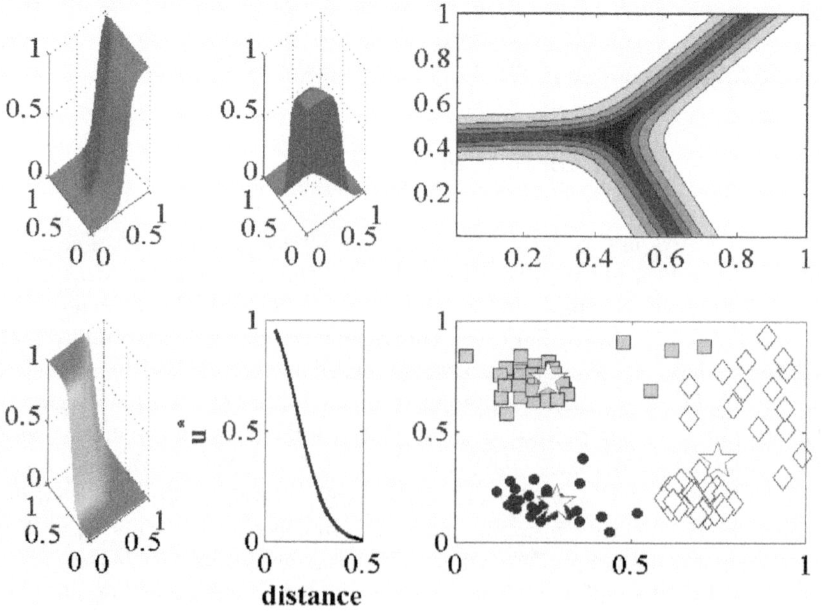

Fig. 6.2. Rather crisply partitioned result by the entropy-based method (**FCM**-e) with $\nu = 0.05$

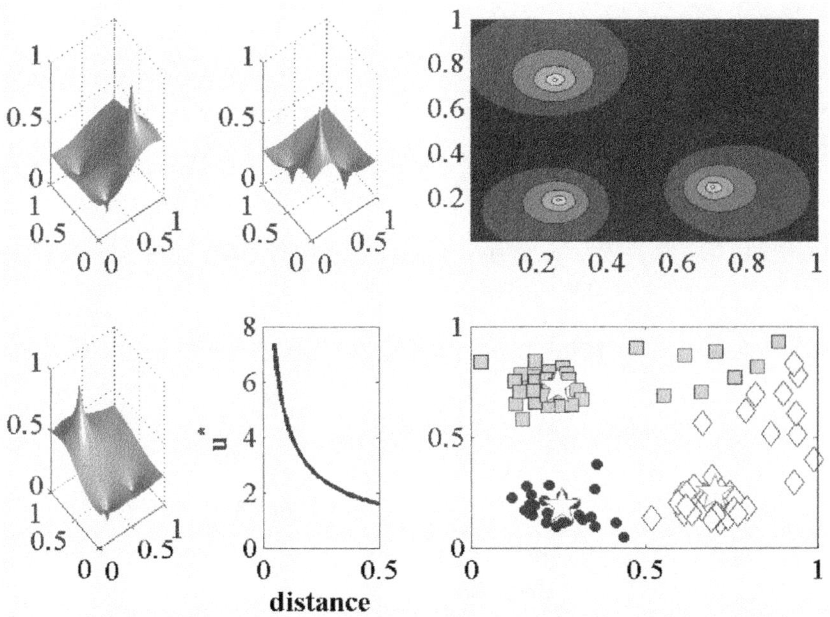

Fig. 6.3. Fuzzily partitioned result by the standard method (**FCM**-s) with $m = 4$

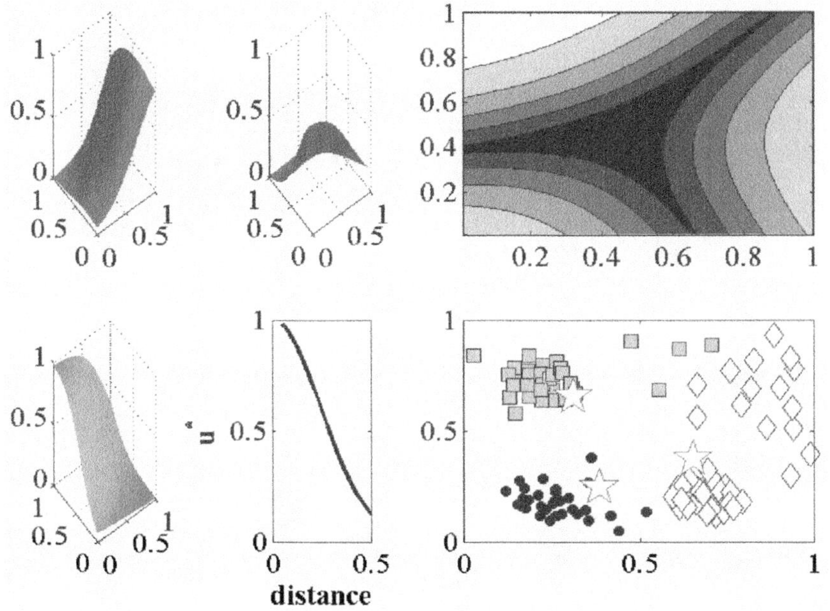

Fig. 6.4. Fuzzily partitioned result by the entropy-based method (**FCM**-e) with $\nu = 0.12$

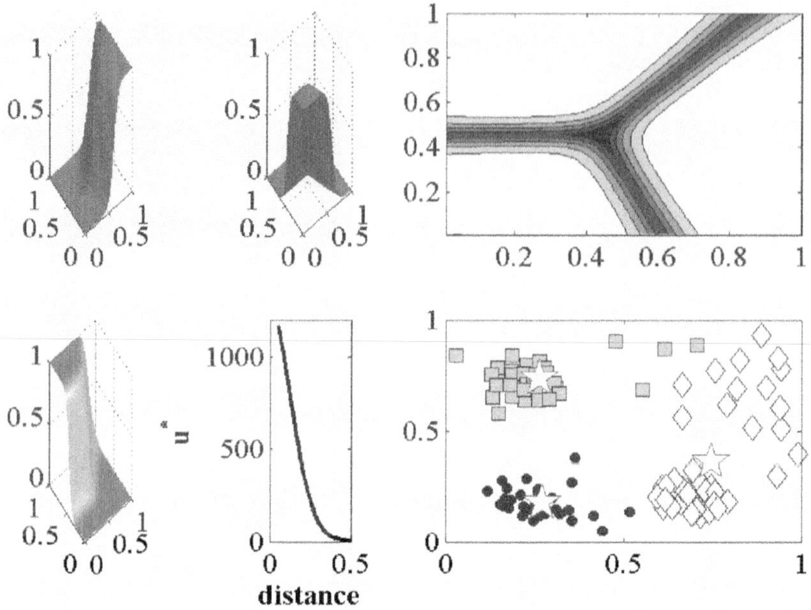

Fig. 6.5. Rather crisply partitioned result by the generalized method (**FCM**-g) with $m = 1.05$ and $\nu = 0.7$

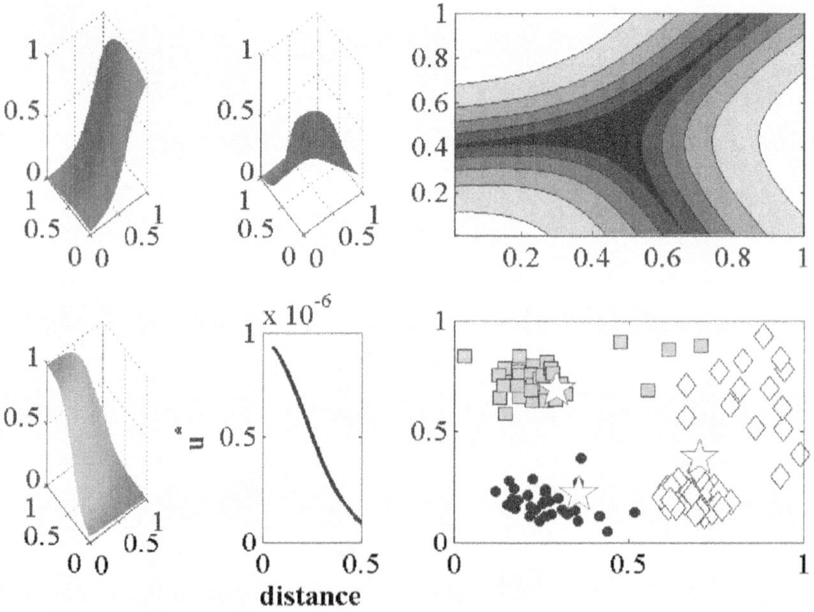

Fig. 6.6. Fuzzily partitioned result by the generalized method (**FCM**-g) with $m = 1.05$ and $\nu = 2$

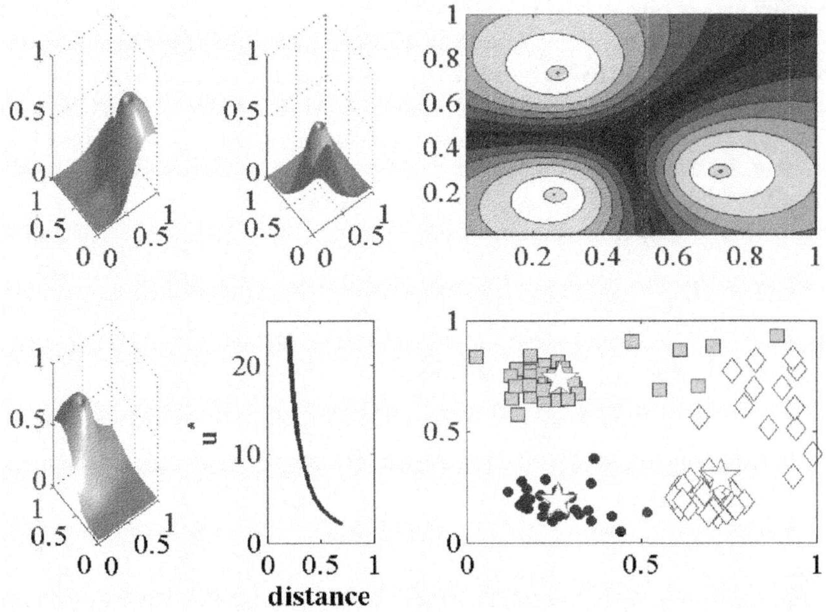

Fig. 6.7. Result by KHM with $m = 1.95$

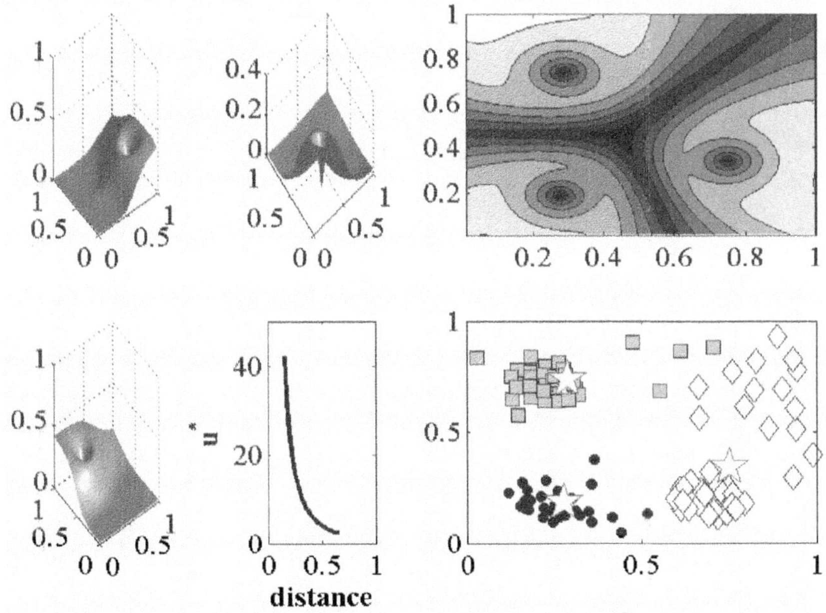

Fig. 6.8. Result by KHM with $m = 1.8$

Fig.6.5 shows rather crisply partitioned result by **FCM**-g with $m = 1.05$ and $\nu = 0.7$. The result is similar to one by **FCM**-e in Fig.6.2. Fig.6.6 shows fuzzily partitioned result by **FCM**-g with $m = 1.05$ and $\nu = 2$. The result also is similar to one by **FCM**-e in Fig.6.4. Since **FCM**-g is reduced to **FCM**-s when the fuzzifier $\nu = 0$, **FCM**-g can produce the same results with those by **FCM**-s. Therefore, **FCM**-g shares the similar clustering characteristics with both **FCM**-s and **FCM**-e.

Figs.6.7 and 6.8 show the clustering result of KHM with $m = 1.9$ and $m = 1.6$ respectively. The 3D graphics shows the weight $\left[\sum_{l=1}^{c}\left(\frac{D(x_k,v_i)}{D(x_k,v_l)}\right)^{\frac{1}{m-1}}\right]^{-2} \times D(x_k,v_i)^{\frac{1}{m-1}-1}$. A dent is seen on each center of the clusters, though the clustering result is similar to one by **FCM**-s.

6.2 Clustering with Iteratively Reweighted Least Square Technique

By replacing the entropy term of the entropy-based method in (6.4) with K-L information term, we can consider the minimization of the following objective function under the constraints that both the sum of u_{ki} and the sum of π_i with respect to i equal one respectively.

$$J_{\mathrm{KL}}(U,V,A,S) = \sum_{i=1}^{c}\sum_{k=1}^{N} u_{ki} D(x_k,v_i;S_i) + \nu \sum_{i=1}^{c}\sum_{k=1}^{N} u_{ki}\log\frac{u_{ki}}{\alpha_i} + \sum_{i=1}^{c}\sum_{k=1}^{N} u_{ki}\log|S_i|, \tag{6.27}$$

where

$$D(x_k,v_i;S_i) = (x_k - v_i)^{\top} S_i^{-1}(x_k - v_i) \tag{6.28}$$

is Mahalanobis distance from x_k to i-th cluster center, and S_i is a covariance matrix of the i-th cluster. From this objective function, we can derive an iterative algorithm of the normal mixture or Gaussian mixture models when $\nu = 2$. From the necessary condition for the optimality of the objective function, we can derive:

$$S_i = \frac{\sum_{k=1}^{N} u_{ki}(x_k - v_i)(x_k - v_i)^{\top}}{\sum_{k=1}^{N} u_{ki}}. \tag{6.29}$$

$$v_i = \frac{\sum_{k=1}^{N} u_{ki} x_k}{\sum_{k=1}^{N} u_{ki}}. \tag{6.30}$$

$$\alpha_i = \frac{\sum_{k=1}^{N} u_{ki}}{\sum_{j=1}^{c}\sum_{k=1}^{N} u_{jk}} = \frac{1}{n}\sum_{k=1}^{N} u_{ki}. \tag{6.31}$$

This is the only case known to date, where covariance matrices (S_i) are taken into account in the objective function $J(U, V)$ in (6.6). Although Gustafson and Kessel's modified **FCM** [45] can treat covariance structures and is derived from an objective function with fuzzifier m, we need to specify the values of determinant $|S_i|$ for all i.

In order to deal with the covariance structure more freely within the scope of **FCM**-g clustering, we need some simplifications based on the iteratively reweighted least square technique. Runkler and Bezdek's [138] fuzzy clustering scheme called alternating cluster estimation (ACE) is this kind of simplification.

Now we consider to deploy a technique from the robust M-estimation [53, 64]. The M-estimators try to reduce the effect of outliers by replacing the squared residuals with ρ-function, which is chosen to be less increasing than square. Instead of solving directly this problem, we can implement it as the iteratively reweighted least square (IRLS). While the IRLS approach does not guarantee the convergence to a global minimum, experimental results have shown reasonable convergence points. If one is concerned about local minima, the algorithm can be run multiple times with different initial conditions.

We implicitly define ρ-function through the weight function. Let us consider a clustering problem whose objective function is written as:

$$J_\rho = \sum_{i=1}^{c}\sum_{k=1}^{N} \rho(d_{ki}). \tag{6.32}$$

where $d_{ki} = \sqrt{D(x_k, v_i; S_i)}$ is a square root of the distance given by (6.28). Let v_i be the parameter vector to be estimated. The M-estimator of v_i based on the function $\rho(d_{ki})$ is the vector which is the solution of the following $m \times c$ equations:

$$\sum_{k=1}^{N} \psi(d_{ki}) \frac{\partial d_{ki}}{\partial v_{ij}} = 0, i = 1, ..., c, j = 1, ..., m \tag{6.33}$$

where the derivative $\psi(z) = d\rho/dz$ is called the influence function. We can define the weight function as:

$$w(z) = \psi(z)/z. \tag{6.34}$$

Since

$$\frac{\partial d_{ki}}{\partial v_i} = -\left((x_k - v_i)^\top S_i^{-1}(x_k - v_i)\right)^{-\frac{1}{2}} S_i^{-1}(x_k - v_i),$$

(6.33) becomes

$$\sum_{k=1}^{N} w(d_{ki}) S_i^{-1}(x_k - v_i) = 0, i = 1, ..., c, \tag{6.35}$$

or equivalently as (6.30), which is exactly the solution to the following IRLS problem. We minimize

$$J_{ifc} = \sum_{i=1}^{c}\sum_{k=1}^{N} w(d_{ki})\left(D(x_k, v_i; S_i) + \log|S_i|\right). \tag{6.36}$$

where we set as $w(d_{ki}) = u_{ki}$. Covariance matrix S_i in (6.29) can be derived by differentiating (6.36). The weight w should be recomputed after each iteration in order to be used in the next iteration. In robust M-estimation, the function $w(d_{ki})$ provides adaptive weighting. The influence from x_k is decreased when $|x_k - v_i|$ is very large and suppressed when it is infinitely large. While we implicitly defined ρ-function and IRLS approach in general does not guarantee the convergence to a global minimum, experimental results have shown reasonable convergence points.

To facilitate competitive movements of cluster centers, we need to define the weight function to be normalized as:

$$u_{ki} = \frac{u^*_{ki}}{\sum_{l=1}^{c} u^*_{lk}}. \tag{6.37}$$

We confine our numerical comparisons to the following two membership functions $u^{*(1)}$ and $u^{*(2)}$ in the next section.

$$u^{*(1)}_{ki} = \frac{\alpha_i |S_i|^{-\frac{1}{\gamma}}}{\left(D(x_k, v_i; S_i)/0.1 + \nu\right)^{\frac{1}{m}}}. \tag{6.38}$$

One popular preprocessing technique is data normalization. Normalization puts the variables in a restricted range with a zero mean and 1 standard deviation. In (6.38), $D(x_k, v_i; S_i)$ is divided by 0.1 so that the proper values of ν are around 1 when all feature values are normalized to zero mean and unit variance, and the data dimensionality is small. This value is not significant and only affects the rage of parameters in searching for proper values.

$$u^{*(2)}_{ki} = \alpha_i |S_i|^{-\frac{1}{\gamma}} \exp\left(-\frac{D(x_k, v_i; S_i)}{\nu}\right). \tag{6.39}$$

Especially for (6.38), u_{ki} of (6.37) can be rewritten as:

$$u_{ki} = \alpha_i |S_i|^{-\frac{1}{\gamma}} \left[\sum_{j=1}^{c} \left(\frac{D(x_k, v_i; S_i)/0.1 + \nu}{D(x_k, v_j; S_i)/0.1 + \nu}\right)^{\frac{1}{m}} \alpha_j |S_j|^{-\frac{1}{\gamma}}\right]^{-1}. \tag{6.40}$$

$u^{*(1)}$ is a modified and parameterized multivariational version of Cauchy's weight function in the M-estimator or of the probability density function (PDF) of Cauchy distribution. It should be noted that in this case (6.37) corresponds to (6.9), but (6.30) is slightly simplified from (6.10). $u^{*(2)}$ is a modified Welsch's weight function in the M-estimator. Both the functions take into account covariance matrices in an analogous manner with the Gaussian PDF. If we choose

$u^{*(2)}$ in (6.39) with $\nu = 2, \gamma = 2$, then **IRLS-FCM** is the same as the Gaussian mixture model (GMM). For any other values of $m > 0$ and $m = \gamma$, **IRLS-FCM** is the same as **FCMAS** whose objective function is $J_{\mathrm{KL}}(U, V, A, S)$ in (6.27).

Algorithm IRLS-FCM: *Procedure of IRLS Fuzzy c-Means.*

IFC1. Generate $c \times N$ initial values for u_{ki} $(i = 1, 2, \ldots, c, k = 1, 2, \ldots, N)$.
IFC2. Calculate $v_i, i = 1, ..., c$, by using (6.30).
IFC3. Calculate S_i and $\alpha_i, i = 1, ..., c$, by using (6.29) and (6.31).
IFC4. Calculate u_{ki}, $i = 1, ..., c, k = 1, ..., n$, by using (6.37) and (6.38).
IFC5. If the objective function (6.36) is convergent then terminate, else go to **IFC2**.
End IFC.

Since $D(x_k, v_i; S_i)$ is Mahalanobis distance, $\sum_{i=1}^{c} \sum_{k=1}^{N} w(d_{ki}) D(x_k, v_i; S_i)$ converges to a constant value, (i.e., the number of sample data $N \times$ the number of variates p).

6.3 FCM Classifier

In the post-supervised classifier design, the clustering is implemented by using the data from one class at a time. The clustering is done on a per class basis. When working with the data class by class, the prototypes (cluster centers) that are found for each labeled class already have the assigned physical labels.

After completing the clustering for all classes, the classification is performed by computing class memberships. Let π_q denote the mixing proportion of class q, i.e., the *a priori* probability of class q. Class membership of k-th data x_k to class q is computed as:

$$u_{qjk}^{*(1)} = \frac{\alpha_{qj} |S_{qj}|^{\frac{1}{\gamma}}}{(D_q(x_k, v_j; S_j)/0.1 + \nu)^{\frac{1}{m}}}, \tag{6.41}$$

$$\tilde{u}_{qk}^{(1)} = \frac{\pi_q \sum_{j=1}^{c} u_{qjk}^{*}}{\sum_{s=1}^{Q} \pi_s \sum_{j=1}^{c} u_{sjk}^{*}}, \tag{6.42}$$

where c denotes the number of clusters of each class and Q denotes the number of classes. The denominator in (6.42) can be disregarded when applied solely for classification. Whereas (6.40) is referred to as a classification function for clustering, (6.42) is a discriminant function for pattern classification. The FCM classifier performs somewhat better than alternative approaches and requires only comparable computation time with Gaussian Mixture classifier because the functional structure of the FCM classifier is similar to that of the Gaussian Mixture classifier. In fact, when $u^{*(2)}$ with $\nu = 2$ and $\gamma = 2$ is used, FCM classifier is the Gaussian Mixture classifier.

The modification of covariance matrices in the mixture of probabilistic principal component analysis (MPCA) [156] or the character recognition [151, 124] is applied in the FCM classifier. P_i is a $p \times p$ matrix of eigenvectors of S_i. p

equals the dimensionality of input samples. Let S_i' denotes an approximation of S_i in (6.29). P_i^r is a $p \times r$ matrix of eigenvectors corresponding to the r largest eigenvalues, where $r < p-1$. Δ_i^r is an $r \times r$ diagonal matrix. r is chosen so that all S_i's are nonsingular and the classifier maximizes its generalization capability.

Inverse of S_i' becomes

$$S_i'^{-1} = P_i^r((\Delta_i^r)^{-1} - \sigma_i^{-1} I_r)P_i^{r\top} + \sigma_i^{-1} I_p, \tag{6.43}$$

$$\sigma_i = (\text{trace}(S_i) - \Sigma_{l=1}^r \delta_{il})/(p-r). \tag{6.44}$$

When r=0, S_i' is reduced to a unit matrix and $D(x_k, v_i; S_i)$ in (6.28) is reduced to Euclidean distance. Then, u_{ki} in (6.40) is reduced to (6.9) when $\alpha_i = 1$ for all i and one is subtracted from m.

Parameter values of m, γ and ν are chosen by optimization methods such as the golden section search method [129] or other recently developed evolutionary algorithms.

6.3.1 Parameter Optimization with CV Protocol and Deterministic Initialization

High performance classifiers usually have parameters to be selected. For example, SVM [162, 19] has the margin and kernel parameters. After selecting the best parameter values by some procedures such as the cross varidation (CV) protocol, we usually fix the parameters and train the whole training set, and then test new unseen data. Therefore, if the performance of classifiers is dependent of random initialization, we need to select parameters with the best average performance and the result of a final single run on the whole training set does not necessarily guarantee the average accuracy. This is a crucial problem and for making our FCM classifier deterministic, we propose a way of determining initial centers based on principal component (PC) basis vectors.

As we will show in the numerical experiment section, the proposed classifier with two clusters for each class (i.e., c=2) performs well, so we let c=2. p_1^* is a PC basis vector of data set $D = (x_1, ..., x_n)^\top$ of a class, which is associated with the largest singular value σ_1^*.

Initial locations of the two cluster centers for the class are given by

$$\begin{aligned} v_1 &= v^* + \sigma_1^* p_1^*, \\ v_2 &= v^* - \sigma_1^* p_1^*, \end{aligned} \tag{6.45}$$

where v^* is the class mean vector. We choose the initial centers in this way, since we know that, for a normal distribution $\mathcal{N}(\mu, \sigma^2)$, the probability of encountering a point outside $\mu \pm 2\sigma$ is 5% and outside $\mu \pm 3\sigma$ is 0.3%.

The FCM classifier has several parameters, whose best values are not known beforehand, consequently some kind of model selection (parameter search) must be done. The goal is to identify good values so that the classifier can accurately predict unseen data (i.e., testing/checking data). Because it may not be useful to achieve high training accuracy (i.e., accuracy on training data whose class labels are known), a common way is to separate training data to two parts of which one is considered unknown in training the classifier. Then the prediction accuracy on this set can more precisely reflect the performance on classifying unknown data. The cross-validation procedure can prevent the overfitting problem. In 10-fold cross-validation (10-CV), we first divide the training set into 10 subsets of equal size. Sequentially one subset is tested using the classifier trained on the remaining 9 subsets. Thus, each instance of the whole training set is predicted once so the cross-validation error rate is the percentage of data which are misclassified. The best setting of the parameters is picked via 10-CV and a recommend procedure is "grid-search". The grid-search is a methodologically simple algorithm and can be easily parallelized while many of advanced methods are iterative processes, e.g. walking along a path, which might be difficult for parallelization. In our proposed approach, the grid-search is applied for m in the unsupervised clustering. We denote this value as m^*.

The golden section search [129] is a technique with iterative processes for finding the extremum (minimum or maximum) of a unimodal function, by successively narrowing brackets by upper bounds and lower bounds. The technique derives its name from the fact that the most efficient bracket ratios are in a golden ratio.

A brief explanation of the procedure is as follows. Let the search interval be the unit interval [0, 1] for the sake of simplicity. If the minimum of a function $f(x)$ lies in the interval, the value of $f(x)$ is evaluated at the three points $x_1 = 0, x_2 = 0.382, x_3 = 1.0$. If $f(x_2)$ is smaller than $f(x_1)$ and $f(x_3)$, we guarantee that a minimum lies inside the interval [0, 1]. The next step in the minimization process is to probe the function by evaluating it at a new point $x_4 = 0.618$ inside the larger interval, i.e. between x_2 and x_3. If $f(x_4) > f(x_2)$ then a minimum lies between x_1 and x_4 and the new triplet of points will be x_1, x_2 and x_4. However if $f(x_4) < f(x_2)$ then a minimum lies between x_2 and x_3, and the new triplet of points will be x_2, x_4 and x_3. Thus, in either case, a new narrower search interval, which is guaranteed to contain the function's minimum, is produced. The algorithm is the repetition of these steps. Let $a = x_2 - x_1, b = x_3 - x_2$ and $c = x_4 - x_2$ then we want:

$$\frac{c}{a} = \frac{a}{b}, \tag{6.46}$$

$$\frac{c}{b - c} = \frac{a}{b}. \tag{6.47}$$

By eliminating c, we have

$$\left(\frac{b}{a}\right)^2 = \frac{b}{a} + 1. \tag{6.48}$$

The solution to (6.48) is the golden ratio $\frac{b}{a} = 1.618033989$... When the search interval is [0, 1], $x_2 = 0.382$ and $x_4 = 0.618$ come from this ratio.

In our proposed approach parameters m, γ and ν are optimized in the post-supervised classification phase by using the golden section search method applied to the parameters one after another. The classification error rate is roughly unimodal with respect to a parameter when other parameters are fixed. The parameters are initialized randomly and updated iteratively, and this procedure is repeated many times, so the procedure can be parallelized.

Our parameter optimization (POPT) algorithm by grid search and golden section search is as follows:

Algorithm POPT: *Procedure of Parameter Optimization.*

POPT1. Initialize $v_i, i = 1, 2$ by using (6.45) and set lower limit (LL) and upper limit (UL) of m^*. Let $m^* = LL$.
POPT2. Partition the training set in 10-CV by **IRLS-FCM** clustering with $\gamma = \nu = 1$. The clustering is done on a per class basis, then all S_i's and v_i's are fixed. Set t:=1.
POPT3. Choose γ and ν randomly from interval [0.1 50].
POPT4. Optimize m for the test set in 10-CV by the golden section search in interval [0.1 1].
POPT5. Optimize γ for the test set in 10-CV by the golden section search in interval [0.1 50].
POPT6. Optimize ν for the test set in 10-CV by the golden section search in interval [0.1 50].
POPT7. If iteration $t < 50$, $t := t + 1$, go to POPT3 else go to POPT8.
POPT8. $m^* := m^* + 0.1$. If $m^* > UL$, terminate else go to POPT2.
End POPT.

In the grid search for m^* and golden section search for m, γ and ν, the best setting of the parameters is picked via 10-fold CV, which minimizes the error rate on test sets. The iteration number for clustering is fixed to 50, which is adequate for the objective function to converge in our experiments. Complete convergence of clustering procedure may not improve the performance, small perturbation of cluster centers may have good effect on the performance. So smaller number of iteration such as 20 may be enough. That depends on data sets.

When r=0, S_i is reduced to a unit matrix and $D(x_k, v_i; S_i)$ in (6.28) is reduced to Euclidean distance. So we change only m by the golden section search method and set $\alpha_{qj} = 1, \pi_q = 1$ for all j and q.

Alternative approaches for parameter optimization are the evolutionary computations such as the genetic algorithm (GA) and the particle swarm optimization (PSO). GA use techniques inspired by evolutionary biology such as inheritance, mutation, selection, and crossover. PSO is a population based stochastic optimization technique inspired by social behavior of bird flocking or fish schooling. PSO shares many similarities with GA.

6.3.2 Imputation of Missing Values

Missing values are common in many real world data sets. The interest in dealing with missing values has continued with the applications to data mining and microarrays [159]. These applications include supervised classification as well as unsupervised classification (clustering).

Usually entire incomplete data samples with missing values are eliminated in preprocessing (the case deletion method). Other well known methods are the mean imputation, median imputation and nearest neighbor imputation [26] procedure. The nearest neighbor algorithm searches through all the dataset looking for the most similar instances. This is a very time consuming process and it can be very critical in data mining where large databases are analyzed.

In the multiple imputation method the missing values in a feature are filled in with values drawn randomly (with replacement) from a fitted distribution for that feature. This procedure is repeated a number of times [92].

In the local principal component analysis (PCA) with clustering [54, 56], not the entire data samples but only the missing values are ignored by multiplying "0" weights over the corresponding reconstruction errors.

Maximum likelihood procedures that use variants of the Expectation-Maximization algorithm can handle parameter estimation in the presence of missing data. These methods are generally superior to case deletion methods, because they utilize all the observed data. However, they suffer from a strict assumption of a model distribution for the variables, such as a multivariate normal model, which has a high sensitivity to outliers.

In this section, we propose an approach to clustering and classification without eliminating or ignoring missing values but with estimating the values. Since we are not concerned about probability distribution such as multivariate normal, without using the terminology "conditional expectation", the estimation is done by the least square method of Mahalanobis distances (6.28). The first term of (6.36) is the weighted sum of Mahalanobis distances between data points and cluster centers. The missing values are some elements of data vector x_k, which are estimated by the least square technique, that is, the missing elements are the solution to the system of linear equations derived from differentiating (6.28) with respect to the missing elements of data vector x_k.

Let $x^i_{kl}, l = 1, ..., p$, be the elements of centered data x^i_k (i.e., $x^i_{kl} = x_{kl} - v_{il}$) and the j-th element x^i_{kj} be a missing element. The objective function for minimizing Mhalanobis distance with respect to the missing value can be written as:

$$L = z^\top S_i^{-1} z - \mu^\top (z - x^i_k), \tag{6.49}$$

where z is the vector of decision variables and μ is the vector of Lagrange multipliers. The elements of μ and x^i_k corresponding to the missing values are zero.

Then, the system of linear equations can be written as:

$$\begin{pmatrix} 2S_i^{-1} & E \\ E & O \end{pmatrix} x^* = b^i_k, \tag{6.50}$$

where

$$E = \mathrm{diag}(1 \cdots 1 \; 0_j \; 1 \cdots 1), \tag{6.51}$$

$$x^* = (z_1 \cdots z_p \; \mu_1 \cdots \mu_{j-1} \; 0 \; \mu_{j+1} \cdots \mu_p), \tag{6.52}$$

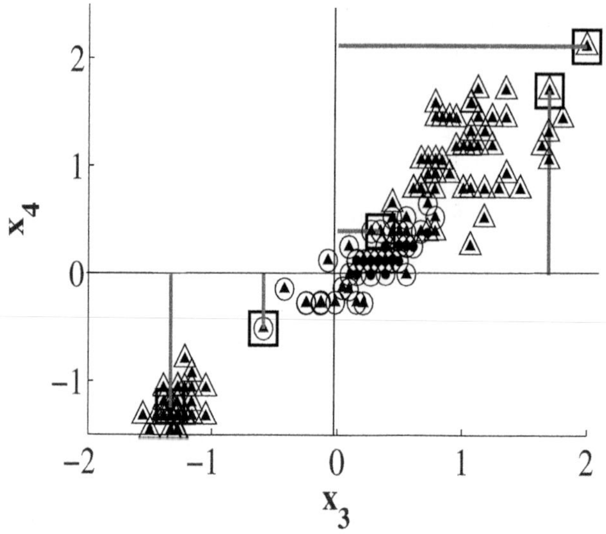

Fig. 6.9. Classification (triangle and circle marks) and imputation (square marks) on Iris 2-class 2D data with missing values by FCM classifier with single cluster for each class ($c = 1$).

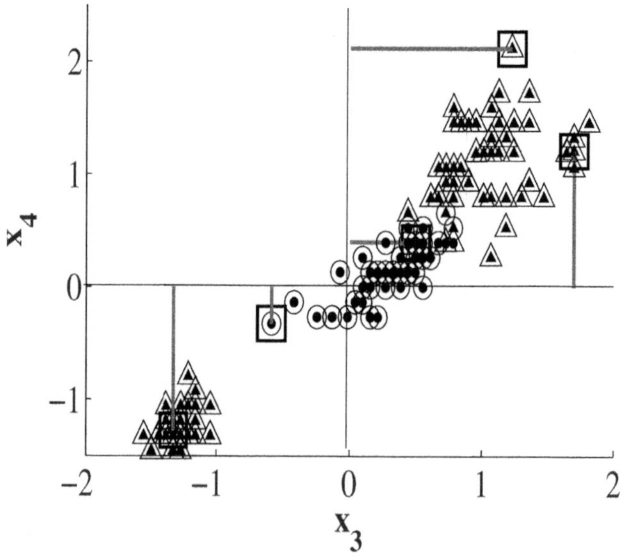

Fig. 6.10. Classification (triangle and circle marks) and imputation (square marks) on Iris 2-class 2D data with missing values by FCM classifier with two clusters for each class ($c = 2$).

and

$$b_k^i = (0_1 \cdots 0_p \; x_{k1}^i \cdots x_{k\;j-1}^i \; 0 \; x_{k\;j+1}^i \cdots x_{kp}^i). \tag{6.53}$$

"diag" denotes diagonal matrix and 0_j denotes that the j-th element is zero. O is a $p \times p$ zero matrix. When more than two elements of x_k are missing, corresponding elements in (6.51)-(6.53) are also replaced by 0. All zero rows and all zero columns are eliminated from (6.50) and then we obtain the least square estimates of all the missing values by adding the element v_{ij} of the i-th center to z_j. We use this estimation method both for clustering and classification. Note that the clustering is done on a per class basis.

Figs.6.9-6.10 show the classification and imputation results on the well known Iris plant data by the FCM classifier with $m = 0.5, \gamma = 2.5, \nu = 1$. Only the two variables, namely x_3 (petal length) and x_4 (petal width) are used and the problem is for binary classification. These figures exemplifies the case where we can easily understand that two clusters are suitable for a class of data with two separate distributions. Open and closed marks of triangle and circle represent the classification decision and the true class (ground truth) respectively. The five artificial data with a missing feature value are added and the values are depicted by solid line segments. Squares mark the estimated values. A vertical line segment shows the x_3 coordinate of an observation for which the x_4 coordinate is missing. A horizontal line segment shows the x_4 coordinate for which the x_3 coordinate is missing. Because the number of clusters is one ($c = 1$) for each class in Fig.6.9, the single class consisted of two subspecies (i.e., Iris Setosa and Iris Verginica) forms a slim and long ellipsoidal cluster. Therefore, the missing values on the right side and on the upper side are estimated at upper right corner of the figure. This problem is alleviated when the number of cluster is increased by one for each class as shown in Fig.6.10.

6.3.3 Numerical Experiments

We used 8 data sets of Iris plant, Wisconsin breast cancer, Ionosphere, Glass, Liver disorder, Pima Indian diabetes, Sonar and Wine as shown in Table 6.2. These data sets are available from the UCI ML repository (http://www.ics.uci.edu/~mlearn/) and were used for comparisons among several prototype-based methods in [165]. Incomplete samples in the breast cancer data set were eliminated from the training and test sets. All categorical attributes were encoded with multi-value (integer) variables. And, then all attribute values were normalized to zero mean and unit variance.

Iris is the set with three classes, though it is known that Iris setosa is clearly separated from the other two subspecies [29] and Iris Versicolor is between Iris Setosa and Iris Verginica in the feature space as shown in Fig. 6.9. If the problem is defined as binary one we can easily find that two clusters are necessary for the Setosa-Verginica class. Iris-Vc and Iris-Vg in the tables represent the Iris subspecies and each of them is two-class binary problem, i.e., one class consists of one subspecies and the other consists of remaining two subspecies. In the same way, Wine-1, Wine-2 and Wine-3 are the binary problems.

Table 6.2. Data sets used in the experiments

	features (p)	objects (N)	classes (Q)
Iris	4	150	3
Breast	9	683	2
Ionosphere	33	351	2
Glass	9	214	6
Liver	6	345	2
Pima	8	768	2
Sonar	60	208	2
Wine	13	178	3

Table 6.3. Classification error rates of **FCMC**, SVM and C4.5 by 10-fold CV with a default partition

	FCMC	SVM	C4.5
Iris	**1.33**	–	5.7 ± 1.3 (4.80)
Iris-Vc	**0.67**	3.33	–
Iris-Vg	**1.33**	2.67	–
Breast	**2.79**	2.65	5.1 ± 0.4 (4.09)
Ionosphere	**3.43**	4.57	–
Glass	**28.10**	–	27.3 ± 1.5 (23.55)
Liver	**27.94**	28.53	33.2 ± 1.4
Pima	**22.50**	22.89	25.0 ± 1.0 (23.63)
Sonar	**10.50**	13.50	24.6 ± 2.7 (19.62)
Wine	**0.00**	–	5.6 ± 1.0
Wine-1	**0.00**	0.59	–
Wine-2	**0.00**	0.59	–
Wine-3	**0.00**	0.59	–

The algorithm was evaluated using 10-CV, which was applied to each data set once for the deterministic classifiers and 10 times for the classifiers with random initializations. We used a default partition into ten subsets of the same size. The performance is the average and standard deviation (s.d.) of classification errors (%).

Table 6.3 shows the results. FCM classifier based on **IRLS-FCM** is abbreviated to **FCMC**. "FCMC" column shows the results by the classifier. POPT algorithm by grid search and golden section search in section 6.3.1 is used. Optimum number (r) of eigenvectors is chosen from 0 (Euclidean distance) up to data dimension. Number of clusters is fixed to two, except for Sonar data where $r = 0$ and Euclidean distance is used. The iteration number for the FCM classifier was fixed to 50, which was adequate in our experiments for the objective function value to converge.

We used downloadable SVM toolbox for MATLAB by A. Schwaighofer (http://ida.first.fraunhofer.de/~anton/software.html) for comparison. The decomposition algorithm is implemented for the training routine, together with efficient working set selection strategies like SVM^{light}[74]. Since SVM is

Table 6.4. Classification error rates of k-NN, LVQ and GMC by 10-fold CV with a default partition

	k-NN	LVQ1	GMC
Iris	2.67	5.40 ± 0.87	2.00 c=1
Iris-Vc	2.67	4.87 ± 0.79	2.80 ± 0.98 c=2
Iris-Vg	2.67	4.40 ± 0.80	4.00 c=1
Breast	2.65	3.16 ± 0.16	2.97 ± 0.13 c=2
Ionosphere	13.43	10.60 ± 0.35	5.71 c=1
Glass	27.62	30.10 ± 1.38	42.38 c=1
Liver	32.65	35.24 ± 1.44	31.68 ± 1.01 c=2
Pima	23.42	24.11 ± 0.52	25.13 c=1
Sonar	13.50	14.85 ± 2.07	17.35 ± 2.19 c=2
Wine	1.76	2.35 ± 0.00	0.59 c=1
Wine-1	1.18	1.59 ± 0.46	0.00 c=1
Wine-2	1.76	3.00 ± 0.49	1.18 c=1
Wine-3	0.59	1.35 ± 0.27	0.00 c=1

Table 6.5. Optimized parameter values used for FCM classifier with two clusters for each class ($c = 2$)

	m^*	m	γ	ν	r	
Iris	0.5	1.0000	8.8014	30.7428	3	
Iris-Vc	0.3	0.1118	0.6177	31.3914	4	
Iris-Vg	0.1	0.8185	0.6177	31.3914	2	
Breast	0.2	0.1000	–	1.0000	0	
Ionosphere	0.6	0.4866	21.7859	49.3433	4	
Glass	1.6	0.1000	2.6307	18.0977	4	
Liver	0.1	0.4703	7.7867	30.9382	4	
Pima	0.7	0.1812	2.0955	17.1634	1	
Sonar	0.2	0.3125	–	1.0000	0	$c = 16$
Wine	0.4	0.9927	16.3793	10.4402	4	
Wine-1	0.1	0.5176	7.3816	10.4402	4	
Wine-2	0.4	0.1000	4.2285	25.5607	4	
Wine-3	0.2	0.1812	4.6000	4.5229	4	

basically for binary classification, we used the binary problems of Iris and Wine. The classification software DTREG (http://www.dtreg.com /index.htm) has a SVM option. The benchmark test results (10-CV) placed on the DTREG web site reports the error rate for the multi-class cases, i.e., 3% on Iris, 34% on Glass, and 1% on Wine.

The third column of Table 6.3 shows the best result among six variants of C4.5 using 10 complete runs of 10-CV reported in [33]. The similar best average error rate among C4.5, Bagging C4.5 and Boosting C4.5 reported in [130] is also displayed in parentheses (standard deviation is not reported).

Nearest neighbor classifier does not abstract the data, but rather uses all training data to label unseen data objects with the same label as the nearest object in the training set. The nearest neighbor classifier easily overfits to the training data. Accordingly, instead of 1-nearest neighbor, generally k nearest neighboring data objects are considered in k-NN classifier. Then, the class label of unseen objects is established by majority vote. For the parameter of k-NN (i.e.,k), we tested all integer values from 1 to 50.

LVQ algorithm we used is LVQ1, which was a top performer (averaged over several data sets) in [9]. Initial value of the learning constant of LVQ was set as 0.3 and was changed as in [9, 165], i.e., $\beta(t+1) = \beta(t) \times 0.8$ where t (=1, ..., 100) denotes iteration number. For the parameter of LVQ (i.e., c), we tested all integer values from 1 to 50.

For GMC, the number of clusters c was chosen from 1 or 2, and optimum number (r) of eigenvectors was chosen similarly with **FCMC**. GMC frequently suffers from the problem of singular matrices and we need to decrease the number of eigenvectors (r) for approximating covariance matrices, though the FCM classifier alleviates the problem. The results are shown in Table 6.4. Table 6.5 shows the parameter values found by the algorithm POPT.

We show in Table 6.6 the results on the benchmark datasets by artificially deleting values. These results were obtained by deleting, at random, observations from a proportion of the instances. The rate of missing feature values with respect to the whole dataset is 25%. From Iris data for example, 150 feature values are randomly deleted. The classification process by 10-CV with a default partition was repeated 10 times for the classifiers with random initializations.

In Table 6.6, "FCMC" column shows the results with the missing value imputation method by the least square Mahalanobis distances. The classification error rates decay only slightly, though the proportion of the instances with missing values is large.

FCMC(M) stands for the FCM classifier with the mean imputation method. Global mean is zero since the data are standardized to zero mean and zero is substituted for the missing value. The proposed **FCMC** is better than the zero imputation method as shown in Table 6.6.

GMC uses EM algorithm and conditional expectation for missing value imputation.

k-NN(NN) is the k-NN classifier with the nearest neighbor imputation method [26]. When the dataset volume is not extremely large, the NN imputation is an efficient method for dealing with missing values in supervised classification. The NN imputation algorithm is as follows:

Algorithm NNI: *Nearest Neighbor Imputation.*

NNI1. Divide the data set D into two parts. Let D_m be the set containing the instances in which at least one of the features is missing. The remaining instances with complete feature information form a set called D_c.

NNI2. Divide the instance vector into observed and missing parts as $x = [x_o; x_m]$.

Table 6.6. Classification error rates (%) on benchmark data sets with artificially deleted values (25%). The results of 10-CV with a default partition.

	FCMC	FCMC(M)	GMC	k-NN(NN)
Iris	**2.67**	4.67	3.33 c=1	9.33
Iris-Vc	**2.67**	5.33	4.67 c=1	9.33
Iris-Vg	**2.67**	6.00	4.67 c=1	6.00
Breast	**3.68**	3.97	3.93 ± 0.11 c=2	4.12
Ionosphere	**4.86**	5.43	7.26 ± 0.68 c=2	–
Glass	**34.76**	35.71	44.29 c=1	48.10
Liver	**33.82**	34.71	39.12 c=1	37.06
Pima	**25.53**	25.00	28.16 c=1	26.05
Sonar	**13.50**	14.00	19.50 c=1	–
Wine	**1.76**	3.53	2.94 c=1	10.59
Wine-1	**0.59**	1.76	3.35 ± 0.95 c=2	4.71
Wine-2	**2.94**	3.53	3.53 c=1	8.82
Wine-3	**0.59**	2.35	1.76 c=1	3.53

Table 6.7. Optimized parameter values used for benchmark data sets with artificially deleted values (25%). Two clusters ($c = 2$) are used in each class.

	m^*	m	γ	ν	r	
Iris	0.8	0.1000	22.4641	10.4402	3	
Iris-Vc	0.4	0.4438	9.5365	13.6211	4	
Iris-Vg	0.6	0.1000	4.2285	25.5607	3	
Breast	0.3	0.1000	–	1.0000	0	
Ionosphere	0.7	0.3390	14.2559	48.9374	4	
Glass	1.9	0.5751	17.2877	37.1569	4	
Liver	0.5	1.0000	4.2285	25.5607	4	
Pima	0.7	0.4319	33.7207	26.3013	1	
Sonar	0.2	0.2313	–	1.0000	0	$c = 16$
Wine	0.6	0.4821	12.2867	26.3013	4	
Wine-1	0.6	0.3863	4.6000	10.4402	4	
Wine-2	0.5	0.1000	2.0535	22.3347	4	
Wine-3	0.1	0.9927	8.0370	13.6211	4	

NNI3. Calculate the distance between the x_o and all the instance vectors from D_c. Use only those features in the instance vectors from the complete set D_c, which are observed in the vector x.
NNI4. Impute missing values from the closest instance vector (nearest neighbor).
End NNI.

Note that all training instances must be stored in computer memory for NN imputation. Sufficient amount of complete data is needed, otherwise it may happen that no complete data exists for substituting the missing value and the computation unexpectedly terminates. k-NN classifier unexpectedly terminated for Ionosphere and Sonar data due to the lack of complete data for nearest neighbor imputation and the result is denoted by "-" in the k-NN(NN) column of Table 6.6.

Tables 6.5 and 6.7 show parameter values used for **FCMC**. Different parameter values are chosen depending on the data sets without or with missing values. When m^* (i.e., fuzzifier) is large, cluster centers come closer each other, so the m^* values determine the position of centers. We see from Tables 6.5 and 6.7, m assumes different values from m^*, that optimizes classifier performance in the post-supervise phase.

Different types of models work best for different types of data, though the FCM classifier outperforms well established classifiers for almost all data sets used in our experiments.

6.4 Receiver Operating Characteristics

Classification performance in terms of miss classification rate or classification accuracy is not the only index for comparing classifiers. The receiver operating characteristics curve, better known as ROC is widely used for diagnosing as well as for judging the discrimination ability of different classifiers. ROC is part of a field called "signal detection theory (SDT)" first developed during World War II for the analysis of radar images. While ROCs have been developed in engineering and psychology [152], they have become very common in medicine, healthcare and particularly in radiology, where it is used to quantify the accuracy of diagnostic test [181].

Determining a proper rejection threshold [36] in classification application is also important in the real-world classification problems, which does not only depend on the actual value of the error rate, but also on the cost involved in taking a wrong decision.

The discriminant analysis based on normal population is a traditional standard approach, which uses posterior probability for classification decisions. Fig.6.11 shows the density functions of normal distributions (upper left) and Cauchy distributions (upper right). Two distributions with equal prior probabilities are drawn and the posterior probability of each distribution computed using the so-called Bayes' rule is shown in the lower row of the figure. Note that in the classification convention the posterior probability is calculated by directly applying the probability density to the Bayes rule. So it is not the actual posterior probability and may be called classification membership. This posterior probability is frequently useful for purposes of identifying the less clear-cut assignment of class membership. A normal distribution in a variate x with mean v and variance σ is a statistic distribution with probability density function (PDF):

$$p(x) = \frac{1}{\sqrt{2\pi}\sigma} \exp\left(-\frac{(x-v)^2}{2\sigma^2} \right). \tag{6.54}$$

While statisticians and mathematicians use the term "normal distribution" or "Gaussian distribution" for this distribution, because of its curved flaring shape, researchers in the fuzzy engineering often refer to it as "bell shape curve." (6.39) is the membership function of multivariate normal type when $\nu = \gamma = 2$.

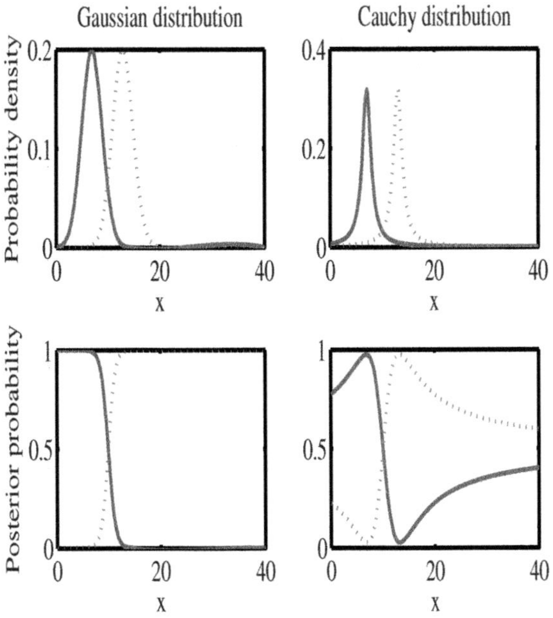

Fig. 6.11. Gaussian distribution with $\sigma = 2$ and Cauchy distribution with $\nu = 1$. Posterior probability means the probability density normalized such that the two density functions sum to one.

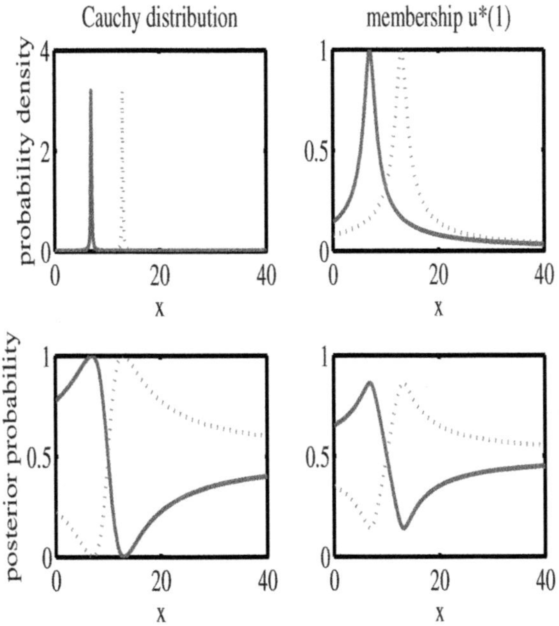

Fig. 6.12. Cauchy distribution with $\nu = 0.1$ and $u^{*(1)}$ with $\nu = 4$

The Cauchy distribution, also called the Lorentzian distribution, is a continuous distribution whose PDF can be written as:

$$p(x) = \frac{1}{\pi} \frac{\nu}{(x - v)^2 + \nu^2}, \tag{6.55}$$

where ν is the half width at half maximum and v is the statistical median. As shown in Fig.6.11 (lower right), posterior probability approaches to 0.5 as x moves to a point distant from v.

Now we define a novel membership function, which is modified from the Cauchy density function, as:

$$u^{(1)}(x) = \frac{1}{((x - v)^2 + \nu)^{\frac{1}{m}}}. \tag{6.56}$$

We call (6.56) as the membership function since we apply it to a fuzzy clustering algorithm. Fig.6.12 (lower right) shows its normalized function with $\nu = 1$.

Though ν of Cauchy distribution is decreased to 0.1 in Fig.6.12, there is no significant difference in its posterior probability. But as shown in Fig.6.12 lower right, the membership function normalized in the same manner as Bayes' rule tends to become close to 0.5 when ν is set to a large value, while the posterior probability of Gaussian distribution approaches zero or one. The property of standard **FCM** membership function is described in Chapter 2. Note that $u^{*(1)}$ defined by (6.56) is a univariate function, whose modified multivariate case is given by (6.38).

In the application to medical diagnosis for example, ROC curves are drawn by classifying patients as diseased or disease-free on the basis of whether they fall above or below a preselected cut-off point which is also referred to as the critical value. The language of SDT is explicitly about accuracy and focuses on two types: sensitivity and specificity. Sensitivity is the proportion of patients with disease whose tests are positive. Specificity is the proportion of patients without disease whose tests are negative. In psychology, these are usually referred to as true positives and correct rejections respectively. The opposite of specificity is the false positive rate or false alarm rate. ROC shows tradeoff between missing positive cases and raising false alarms. A ROC curve demonstrates several things.

1) the closer the curve approaches the top left-hand corner of the plot, the more accurate the test;
2) the closer the curve is to a 45° diagonal, the worse the test (random performance);
3) the area under the curve is a measure of the accuracy of the test;
4) the plot highlights the trade-off between the true positive rate and the false positive rate: an increase in true positive rate is accompanied by an increase in false positive rate.

The problem of determining a proper rejection threshold [36] in classification application is also a major topic in the process of development of real-world

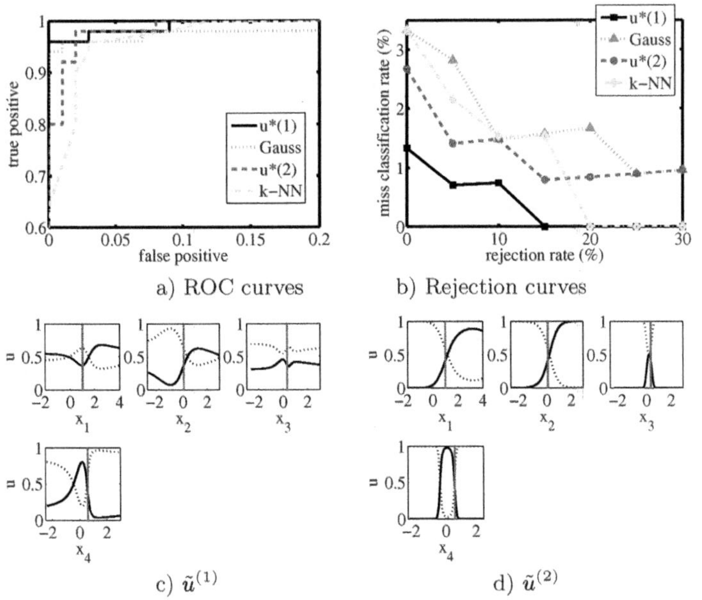

Fig. 6.13. Results on the iris Versicolor data. a) ROC curves, b) rejection curves, c) $\tilde{u}^{(1)}$ and d) $\tilde{u}^{(2)}$. $u^{*(1)}: c = 2, m = 0.6, \gamma = 3, \nu = 1, p = 4$, $u^{*(2)}$(Gauss): $c = 1, \nu = 2, \gamma = 2, p = 2$, $u^{*(2)}: c = 2, \nu = 1.7, \gamma = 1.7, p = 4$, k-NN: k=13.

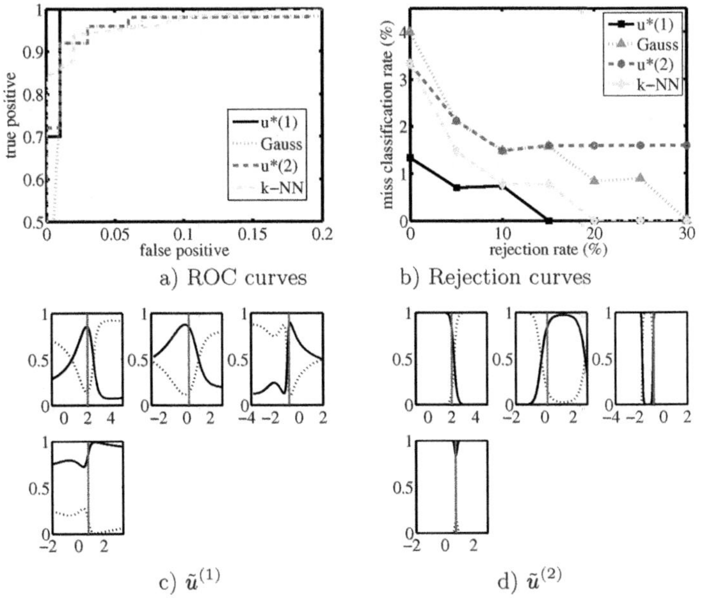

Fig. 6.14. Results on the iris Verginica data. a) ROC curves, b) rejection curves, c) $\tilde{u}^{(1)}$ and d) $\tilde{u}^{(2)}$. $u^{*(1)}: c = 2, m = 0.5, \gamma = 2.4, \nu = 1, p = 4$, $u^{*(2)}$(Gauss): $c = 1, \nu = 2, \gamma = 2, p = 3$, $u^{*(2)}: c = 1, \nu = 0.5, \gamma = 2, p = 3$, k-NN: k=8.

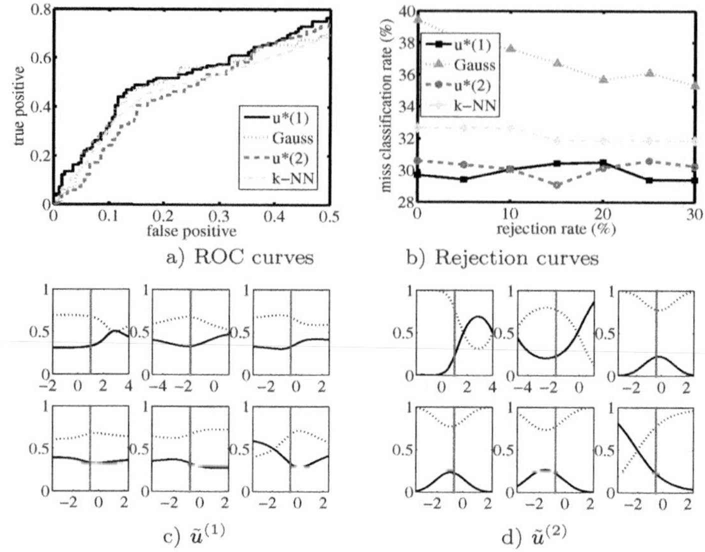

Fig. 6.15. Results on the liver disorder data. a) ROC curves, b) rejection curves, d) $\tilde{u}^{(1)}$ and e) $\tilde{u}^{(2)}$. $u^{*(1)}: c = 1, m = 1, \gamma = 7, \nu = 1, p = 6$, $u^{*(2)}$(Gauss): $c = 1, \nu = 2, \gamma = 2, p = 2$, $u^{*(2)}: c = 1, \nu = 0.5, \gamma = 11, p = 5$, k-NN: k=23.

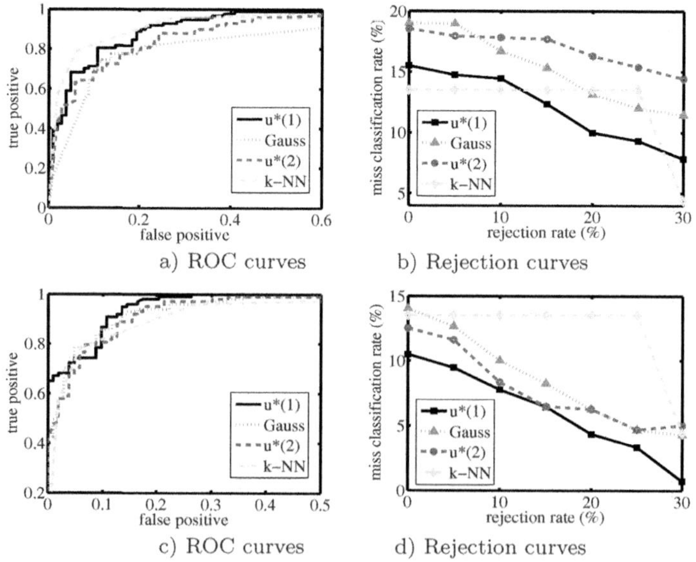

Fig. 6.16. Results on the sonar data. a) ROC curves and b) rejection curves with $u^{*(1)}: c = 1, m = 0.5, \gamma = 16, \nu = 1, p = 8$, $u^{*(2)}$(Gauss): $c = 1, p = 7$, $u^{*(2)}: c = 1, \nu = 3, \gamma = 10, p = 36$, k-NN: k=3. c) ROC curves and d) rejection curves with $u^{*(1)}: S = I, c = 20, \nu = 0.2$, $u^{*(2)}$(Gauss): $S = I, c = 30, \nu = 2$ and $u^{*(2)}: S = I, c = 20, \nu = 2.2$, k-NN: k=3.

recognition systems. There are practical situations in which, if the miss classification rate is too high, it could be too risky to follow the decision rule, and avoiding to classify the current observation could be better. In the medical application for example, additional medical tests would be required for a determinate diagnosis. The decision does not only depend on the actual value of the error rate, but also on the cost involved in taking a wrong decision. Confidence values are used as rejection criterion. Samples with a low confidence can be rejected in order to improve the quality of the classifier. It is expected that when rejecting samples the classification performance will increase, although it may be the case that false rejections are done. On the average, however, it is expected that the samples are correctly rejected. We are able to compare the classifier performance by drawing rejection curves.

Fig.6.13-6.16 show the ROC and rejection curves. The sub figures in the lower column show the class membership functions $\tilde{u}^{(1)}$ (lower left) and $\tilde{u}^{(2)}$ (lower right), which are classification functions and are normalized such that those of the two classes sum to one as in (6.42). The membership functions at the level of a sample data (depicted by a vertical line at the center) on each of the attributes, i.e., variates $x_1, x_2, ..., x_p$, are shown from left to right and top to bottom. "Gauss" represents the Gaussian classifier based on normal populations, which is equivalent to the FCM classifier by $u^{*(2)}$ with $c = 1, \gamma = 2$ and $\nu = 2$. The parameter values are described in the legend of the figures. They may not be completely global optimal but approximately optimal. The optimal integers are chosen for the parameter k of k-NN. The true positive rate, false positive rate and misclassification rate are on the test sets by 10-fold CV.

Fig.6.13 shows the results on the Iris-Vc data. In Fig.6.13, a) shows ROC curves. True positive rates are plotted against false positive rates. The closer the ROC curve approaches the top left-hand corner of the plot, the more accurate the test. The curve of the FCM classifier using $u^{*(1)}$ is plotted by straight lines, which are close to the top left-hand corner. Fig.6.13 b) shows the rejection curves. c) and d) respectively show $\tilde{u}^{(1)}$ and $\tilde{u}^{(2)}$ on each of the four variables. Fig.6.14 shows the results on the Iris-Vg data.

The liver disorder data and Pima Indian diabetes data are difficult to discriminate between the two classes and the miss classification rates are greater than 20%. As shown in Fig.6.15 c), the membership or classification function takes values close to 0.5, which signify the small confidence in the classification decision of the liver data. The smaller confidence seems to be more rational for a dataset whose classification accuracy is low.

In Fig.6.16, the results on the sonar data when c=1 are compared to those with c=20 or 30. Euclidean distance is used instead of Mahalanobis distance. With many clusters and Euclidian distance, the miss classification rates are decreased for both $u^{*(1)}$ and $u^{*(2)}$. Since the sonar data have 33 feature variables, the membership functions are not plotted in Figs.6.16.

The classification performances in terms of ROC and rejection largely depend on the data sets.

6.5 Fuzzy Classifier with Crisp c-Means Clustering

In supervised classifier design, a data set is usually partitioned into a training set and a test set. Testing a classifier designed with the training set means finding its misclassification rate. The standard method for doing this is to submit the test set to the classifier and count errors. This yields the performance index by which the classifier is judged because it measures the extent to which the classifier generalizes to the test data. When the test set is equal to the training set, the error rate is called the resubstitution error rate. This error rate is not reliable for assessing the generalization ability of the classifier, but this is not an impediment to using as a basis for comparison of different designs. If training set is large enough and its substructure is well delineated, we expect classifiers trained with it to yield good generalization ability. It is not very easy to choose the classifier or its parameters when applying to real classification problems, because the best classifier for the test set is not necessarily the best for the training set. While the FCM classifier in Chapter 6.3 is designed to maximize the accuracy for test set, the fuzzy classifier with crisp c-means clustering (**FCCC**) in this section is designed to maximize the accuracy for training set and we confine our comparisons to the resubstitution classification error rate and the data set compression ratio as performance criteria.

6.5.1 Crisp Clustering and Post-supervising

In Section 3.5, the generalized crisp clustering is derived from the objective function $J_{\mathrm{dKL}}(U, V, A, S)$ by setting $\nu = 0$ [113]. $J_{\mathrm{dKL}}(U, V, A, S)$ is a defuzzified clustering objective functional. Similarly, we can derive the same crisp clustering algorithm from (6.40), and we include it in the family of **CMO** in Chapter 2.

FCMC in Section 6.3 is a fuzzy approach and post-supervised, and the IRLS clustering phase can be replaced by a crisp clustering algorithm. Although the crisp clustering algorithm in Section 3.5.2 is the sequential crisp clustering algorithm, for simplicity's sake we confine our discussion to its simple batch algorithm. Note that the **CMO** algorithm in Chapter 2 uses S_i of unit matrix, and $D(x_k, v_i; S_i)$ in (6.28) is Euclidean distance.

An alternating optimization algorithm is the repetition of (6.29) through (6.31) and

$$u_{ki} = \begin{cases} 1 & (i = \arg\min_{1 \le j \le c} D(x_k, v_j; S_j) + \log|S_j|), \\ 0 & (\text{otherwise}). \end{cases} \tag{6.57}$$

The modification of covariance matrices by (6.43) is not enough for preventing singular matrices when the number of instances included in a cluster is very small

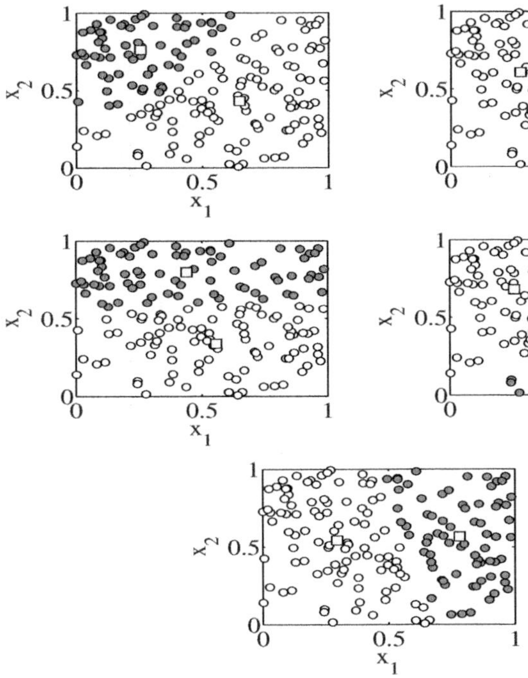

Fig. 6.17. Five different clustering results observed by 10 trials of **CMO**

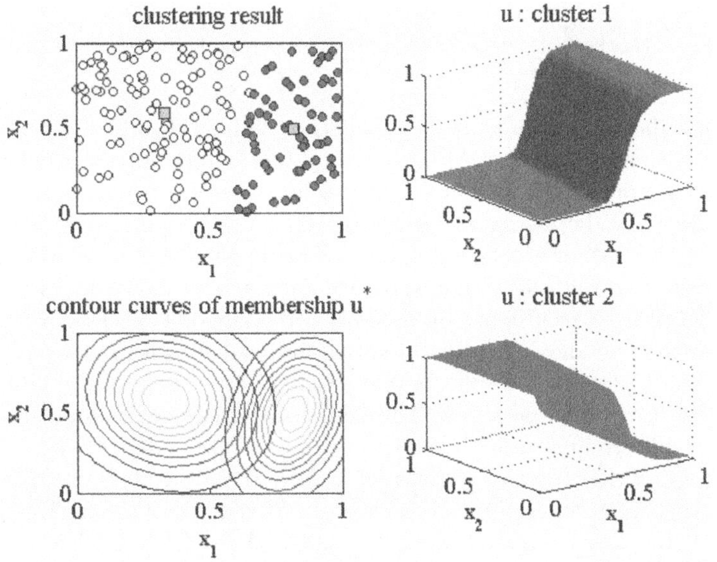

Fig. 6.18. Result observed 9 times out of 10 trials by GMM

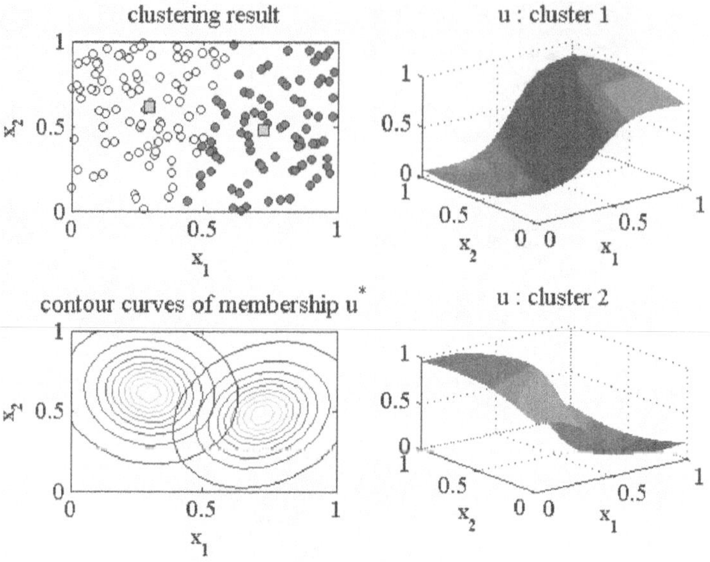

Fig. 6.19. Result observed 9 times out of 10 trials by **IRLS-FCM**

or zero. When the number becomes too small and an S_i results in a singular matrix, for increasing the number we modify (6.57) as:

$$u_{ki} = \begin{cases} 0.9 & (i = \arg\min_{1 \le j \le c} D(x_k, v_j; S_j) + \log|S_j|), \\ \frac{0.1}{c-1} & (\text{otherwise}). \end{cases} \tag{6.58}$$

By this fuzzification of membership, even the smallest cluster may include some instances with small membership values, and the centers come somewhat near to the center of gravity of each class. This fuzzification works out, though is being ad hoc. We show the results in section 6.5.2.

Figs.6.17-6.18 show clustering results of artificial 2-D data. **CMO** produces many different results for a non-separate data set as shown in Fig.6.17. Five different clustering results are obtained by 10 trials of **CMO**, while the result similar to the one shown in Fig.6.18 was obtained 9 times out of 10 trials by GMM. Fig.6.19 show the result obtained 9 times out of 10 trials by **IRLS-FCM** with $m = 0.6, \gamma = 0.5$ and $\nu = 1$. The hard clustering algorithm produces many kinds of results compared with the Gaussian and fuzzy clusterings.

As we apply the classifier to data with more than one class, we usually have many more local minima of the clustering criterion of **CMO**. Convergence speed by **CMO** was much faster than GMM and **IRLS-FCM**. **CMO** needed only around 10 iterations, while GMM and **IRLS-FCM** needed more than 50 iterations.

Our proposed classifier is of post-supervised and, thus, the optimum clustering result with respect to the objective function does not guarantee the minimum classification error. Our strategy is to select the best one in terms of classification error from many local minima of the clustering criterion of **CMO**. Parameter values used for **FCCC** are chosen by the golden section search [129], which is applied to m, γ and ν one after another with random initializations.

FCCC algorithm with golden section search method used in the next section is as follows:

Algorithm FCCC: *Procedure of Fuzzy Classifier with Crisp Clustering.*

FCCC1. Initialize v_i's by choosing data vectors randomly from each class.
FCCC2. Partition the training set by **CMO** and fix S_i and v_i.
FCCC3. Choose γ and ν randomly from interval [0.01 5].
FCCC4. Optimize m by the golden section search in interval [0.01 5].
FCCC5. Optimize γ by the golden section search in interval [0.01 5].
FCCC6. Optimize ν by the golden section search in interval [0.01 5].
FCCC7. If iteration $t < 500$, $t := t + 1$, go to FCCC1 else terminate.
End FCCC.

6.5.2 Numerical Experiments

In Table 6.8, "FCCC" column shows the best resubstitution error rates on training sets from 500 trials of clustering by **CMO** and the golden section search. "M" and "E" indicate that Mahalanobis and Euclid distances are used respectively. LVQ result is also the best one from 500 trials on each set with random initializations. Initial value of LVQ learning rate β was set as 0.3 and was changed as in [165], i.e., $\beta(t+1) = \beta(t) \times 0.8$ where t (=1, ..., 100) denotes iteration number. The resubstitution error rates of **FCMC** is the best results from 10 runs of clustering with different m^* and 50 runs of the golden section search on each clustering result. Since **FCMC** uses **IRLS-FCM**, which is not a crisp clustering, 10 runs of clustering seem enough.

For $c > 2$, we set as $p = 0$, then **CMO** is the crisp c-means clustering with Euclidean distances. Naturally, as the number c is increased, the resubstitution

Table 6.8. Best resubstitution error rates from 500 trials by **FCMC**, **FCCC** and LVQ

	FCMC	FCCC			LVQ1	
	M	M	E	E	E	E
	c=2	c=2	c=5	c=10	c=5	c=10
Iris	0.67	**0.67**	1.33	0.67	2.0	1.33
Breast	1.90	**1.76**	2.78	2.05	2.34	1.61
Ionosphere	3.13	**0.85**	5.13	3.70	7.41	5.70
Glass	18.69	**9.81**	18.69	13.08	20.56	18.22
Liver	23.19	**18.84**	25.22	23.48	27.54	21.45
Pima	20.18	**19.40**	20.44	19.14	20.18	18.88
Sonar	5.29	**0**	6.25	0.48	4.81	1.92
Wine	0.00	**0**	0	0	0	0

Table 6.9. Parameter values used for **FCCC** with Mahalanobis distances (c=2)

	r	m	γ	ν
Iris	1	0.29	3.31	4.96
Breast	7	0.05	0.78	2.26
Ionosphere	20	0.04	1.29	1.96
Glass	9	0.01	0.25	4.96
Liver	4	0.27	4.06	2.82
Pima	4	0.16	1.90	4.36
Sonar	9	0.03	1.16	4.96
Wine	1	0.04	1.21	4.96

Table 6.10. Parameter values used for **FCCC** with Euclidean distances (c=5, 10)

	c=5, r=0		c=10, r=0	
	m	ν	m	ν
Iris	1.11	0.01	1.08	0.01
Breast	1.16	1.92	1.04	1.92
Ionosphere	1.02	1.92	1.08	1.92
Glass	1.43	0.01	1.15	1.92
Liver	4.85	5.00	1.19	1.92
Pima	1.49	3.82	1.89	1.19
Sonar	1.02	1.92	1.04	1.92
Wine	1.08	1.92	1.23	1.92

Table 6.11. Compression ratios (%)

	M c=2	E c=5	E c=10
Iris	8.0	10	20
Breast	4.7	1.5	2.9
Ionosphere	23.9	2.8	5.7
Glass	56.1	14.0	28.0
Liver	5.8	2.9	5.8
Pima	2.6	1.3	2.6
Sonar	19.2	4.8	9.6
Wine	6.7	8.42	16.9

error rate decreases and for example when $c = 50$ the rate is 1.17% for Breast cancer data. Since Glass data have 6 classes, when c=2 and (6.57) is used, all trials results in a singular covariance matrix and unexpectedly terminate due to the lack of instances. By using (6.58) the algorithm successfully converged.

Despite the continuous increase in computer memory capacity and CPU speed, especially in data mining, storage and efficiency issues become even more and more prevalent. For this reason we also measured the compression ratios of the trained classifiers in Table 6.11. The ratio is defined as $Ratio=(r+1)\times c\times$ number of classes $\div$ number of instances. The ratios for **FCCC** ($c > 2$) and LVQ are the same. For **FCCC** with Mahalanobis distances and c=2, the compression ratios of Ionosphere and Glass are high, though the error rate is small in Table 6.8. When r=3 and c=2, the best error rate for Ionosphere is 2.85% and the compression ratio is 4.56%. The error rate for the glass data is 10.28% and the compression ratio is 33.6% when r=5 and c=2. **FCCC** demonstrates relatively low compression ratios. Parameter values of **FCCC** chosen by the golden section search method are shown in Table 6.9. **FCCC** with Mahalanobis distance and c=2 attains the lowest error rate when $c \leq 5$ as indicated by boldface letters in Table 6.8. The compression ratios of **FCCC** is not so good for Ionosphere, Glass and Sonar, though we can conjecture from the results of **FCMC** that the generalization ability will not deteriorate largely since only two clusters for each class are used.

7 Fuzzy Clustering and Probabilistic PCA Model

Fuzzy clustering algorithms have close relationship with other statistical techniques. In this chapter, we describe the connection between fuzzy clustering and related techniques.

7.1 Gaussian Mixture Models and FCM-type Fuzzy Clustering

7.1.1 Gaussian Mixture Models

We first give a brief review of the EM algorithm with mixture of normal densities. Density Estimation is a fundamental technique for revealing intrinsic structure of data sets and tries to estimate the most likely density functions. For multi-modal densities, mixture models of simple uni-modal densities are often used.

Let $X = \{x_1, \ldots, x_N\}$ denote p dimensional observations of N samples and Φ be the vector of parameters $\Phi = (\alpha_1, \ldots, \alpha_c, \phi_1, \ldots, \phi_c)$. The mixture density for a sample x is given as the following probability density function:

$$p(x|\Phi) = \sum_{i=1}^{c} \alpha_i p_i(x|\phi_i), \tag{7.1}$$

where conditional density $p_i(x|\phi_i)$ is the component density and mixing coefficient α_i is the a priori probability. In the mixture of normal densities or the Gaussian mixture models (GMMs, e.g., [10, 29]), the component densities are Gaussian distributions with a vector parameter $\phi_i = (\mu_i, \Sigma_i)$ composed of the covariance matrix Σ_i and the mean vector μ_i:

$$p_i(x|\phi_i) \sim \mathcal{N}(\mu_i, \Sigma_i), \tag{7.2}$$

where Σ_i is chosen to be full, diagonal or spherical (a multiple of the identity). The density functions are derived as the maximum likelihood estimators where the log-likelihood to be maximized is defined as

$$L(\Phi) = \sum_{k=1}^{N} \log\left\{ \sum_{i=1}^{c} \alpha_i p_i(x_k|\phi_i) \right\}. \tag{7.3}$$

S. Miyamoto et al.: Algorithms for Fuzzy Clustering, STUDFUZZ 229, pp. 157–169, 2008.
springerlink.com

For searching the maximum likelihood estimators, we can use the expectation-maximization (EM) algorithm. The iterative algorithm is composed of E-step (Expectation step) and M-step (Maximization step), in which $L(\Phi)$ is maximized by maximizing the conditional expectation of the complete-data log-likelihood given a previous current estimate $\hat{\Phi}$ and $x_k, k = 1, \ldots, N$:

$$Q(\Phi|\hat{\Phi}) = \sum_{i=1}^{c}\sum_{k=1}^{N} \psi_{ik} \left\{\log \alpha_i + \log p_i(x_k|\phi_i)\right\}. \tag{7.4}$$

Here, we consider the updating formulas in the two steps in the case of full covariance matrices.

The EM algorithm

(O) Set an initial value $\Phi^{(0)}$ for the estimate. Let $\ell = 0$. Repeat the following **(E)** and **(M)** until convergence.

(E) (Expectation)
Calculate $Q(\Phi|\Phi^{(\ell)})$ by estimating the responsibility (posterior probability) of each data sample for component densities:

$$\begin{aligned} \psi_{ik} &= \frac{\alpha_i p_i(x_k|\phi_i)}{\sum_{j=1}^{c} \alpha_j p_j(x_k|\phi_j)} \\ &= \frac{\alpha_i \exp\left(-\frac{1}{2}E(x_k, \mu_i; \Sigma_i)\right)|\Sigma_i|^{-\frac{1}{2}}}{\sum_{j=1}^{c} \alpha_j \exp\left(-\frac{1}{2}E(x_k, \mu_j; \Sigma_j)\right)|\Sigma_j|^{-\frac{1}{2}}}, \end{aligned} \tag{7.5}$$

where

$$E(x_k, \mu_i; \Sigma_i) = (x_k - \mu_i)^\top \Sigma_i^{-1}(x_k - \mu_i). \tag{7.6}$$

(M) (Maximization)
Find the maximizing solution

$$\bar{\Phi} = \arg\max_{\Phi} Q(\Phi|\Phi^{(\ell)})$$

by updating the parameters of Gaussian components:

$$\alpha_i = \frac{1}{N}\sum_{k=1}^{N} \psi_{ik}, \tag{7.7}$$

$$\mu_i = \frac{\sum_{k=1}^{N} \psi_{ik} x_k}{\sum_{k=1}^{N} \psi_{ik}}, \tag{7.8}$$

$$\Sigma_i = \frac{\sum_{k=1}^{N} \psi_{ik}(x_k - \mu_i)(x_k - \mu_i)^\top}{\sum_{k=1}^{N} \psi_{ik}}. \tag{7.9}$$

Let $\ell := \ell + 1$ and $\Phi^{(\ell)} = \bar{\Phi}$.

7.1.2 Another Interpretation of Mixture Models

Next, we discuss the classification aspect of GMMs. Hathaway [47] gave another interpretation of the optimization procedure. While the EM algorithm is a practical technique for calculating the maximum-likelihood estimates for certain mixtures of distributions, the framework can be regarded as a method of coordinate descent on a particular objective function. From classification view point, the log-likelihood function of (7.4) can be decomposed into two components as follows:

$$\begin{aligned} Q(\Phi|\hat{\Phi}) = & \sum_{i=1}^{c}\sum_{k=1}^{N} \psi_{ik} \log\left(\alpha_i p_i(x_k|\phi_i)\right) \\ & - \sum_{i=1}^{c}\sum_{k=1}^{N} \psi_{ik} \log \psi_{ik} \end{aligned} \tag{7.10}$$

The first term is a classification criterion of a weighted distances function. Indeed, for GMMs with equal proportions and a common spherical covariance matrix, the first term of (7.10) is equivalent to the clustering criterion of the k-means (hard c-means) algorithm. For more general cases, the criterion is identified with the classification maximum likelihood (CML) criterion [17], in which $\psi_k = (\psi_{1k}, \ldots, \psi_{ck})$ is the indicator vector that identifies the mixture component origin of data sample k. The classification EM algorithm (CEM algorithm) [16] is a classification version of the EM algorithm including an additional step of a "C-step" between the E-step and the M-step of the EM algorithm. The C-step converts ψ_{ik} to a discrete classification (0 or 1) based on a maximum a posteriori principle. Therefore, the CEM algorithm performs hard classification.

In the Hathaway's interpretation, the second term of (7.10) is regarded as a penalty term. The sum of entropies [6] is a measure of statistical uncertainty of partition given by the posterior probabilities ψ_{ik}, and is maximized when $\psi_{ik} = 1/c$ for all i and k. Then, the penalty term tends to pull ψ_{ik} away from the extremal values (0 or 1). In this context, the EM algorithm for GMMs is a penalized version of hard c-means, which performs an alternated maximization of $L(\Phi)$ where both alternated steps are precisely the E-step and the M-step of the EM algorithm.

In the penalized hard means clustering model, the calculation of the maximum-likelihood estimation of posterior probabilities ψ_{ik} can be regarded as the minimization of the sum of Kullback-Leibler information (KL information). Equation (7.4) is also decomposed as

$$Q(\Phi|\hat{\Phi}) = -\sum_{k=1}^{N}\Big(\sum_{i=1}^{c}\psi_{ik}\log\big(\psi_{ik}/[\alpha_i p_i(x_k|\phi_i)/p(x_k)]\big)\Big) + \sum_{k=1}^{N}\log p(x_k), \tag{7.11}$$

and the KL information term $\psi_{ik}\log\big(\psi_{ik}/[\alpha_i p_i(x_k|\phi_i)/p(x_k)]\big)$ is regarded as the measure of the difference between ψ_{ik} and $\alpha_i p_i(x_k|\phi_i)/p(x_k)$. Then, the classification parameter ψ_{ik} is estimated so as to minimize the difference between them. In this way, probabilistic mixture models are a statistical formulation for data clustering incorporating with soft classification.

7.1.3 FCM-type Counterpart of Gaussian Mixture Models

When we use the KL information based method [68, 69], the **FCMAS** algorithm can be regarded as an FCM-type counterpart of the GMMs with full unknown parameters. The objective function is defined as follows:

$$J_{\mathrm{klfcm}}(U,V,A,S) = \sum_{i=1}^{c}\sum_{k=1}^{N}u_{ki}D(x_k,v_i;S_i) + \nu\sum_{i=1}^{c}\sum_{k=1}^{N}u_{ki}\log\frac{u_{ki}}{\alpha_i} + \sum_{i=1}^{c}\sum_{k=1}^{N}u_{ki}\log|S_i|, \tag{7.12}$$

where A and S are the total sets of α_i and S_i, respectively. The constraint for A is

$$\mathcal{A} = \{\ A = (\alpha_1,\ldots,\alpha_c)\ :\ \sum_{j=1}^{c}\alpha_j = 1\ ;\ \alpha_i \geq 0,\ 1 \leq i \leq c\ \}. \tag{7.13}$$

$D(x_k, v_i; S_i)$ is the Mahalanobis distance

$$D(x_k,v_i;S_i) = (x_k - v_i)^{\top}S_i^{-1}(x_k - v_i), \tag{7.14}$$

and all the elements of S_i are also decision variables. Equation (7.12) is minimized under the condition that both the sum of u_{ki} and the sum of α_i with respect to i equal 1, respectively. As the entropy term in the entropy based clustering method forces memberships u_{ki} to take similar values, the KL information term of (7.12) is minimized if u_{ki}, $k = 1,\ldots,N$ take the same value α_i within cluster i for all k. If $u_{ki} \simeq \alpha_i$ for all i and k, the KL information term becomes 0 and the membership assignment is very fuzzy; but when ν is 0 the optimization problem reduces to a linear one, and the solution u_{ki} are obtained at the extremal point (0 or 1). Fuzziness of the partition can be controlled by ν. From the necessary conditions, the updating rules in the fix-point iteration algorithm are given as follows:

$$u_{ki} = \frac{\alpha_i \exp\left(-\frac{1}{\nu} D(x_k, v_i; S_i)\right) |S_i|^{-\frac{1}{\nu}}}{\sum_{j=1}^{c} \alpha_j \exp\left(-\frac{1}{\nu} D(x_k, v_j; S_j)\right) |S_j|^{-\frac{1}{\nu}}}, \tag{7.15}$$

$$v_i = \frac{\sum_{k=1}^{N} u_{ki} x_k}{\sum_{k=1}^{N} u_{ki}}, \tag{7.16}$$

$$\alpha_i = \frac{1}{N} \sum_{k=1}^{N} u_{ki}, \tag{7.17}$$

$$S_i = \frac{\sum_{k=1}^{N} u_{ki}(x_k - v_i)(x_k - v_i)^\top}{\sum_{k=1}^{N} u_{ki}}. \tag{7.18}$$

In the KL information based method, KL information term is used for both optimization of cluster capacities and fuzzification of memberships while Hathaway [47] interpreted the clustering criterion as the sum of KL information for updating memberships.

Considering the close relationships between fuzzy clustering techniques and probability models indicate, several extended versions of probabilistic models were proposed incorporating with fuzzy approaches. Gath and Geva [39] proposed the combination of the **FCM** algorithm with maximum-likelihood estimation of the parameters of the components of mixtures of normal distributions, and showed that the algorithm is more robust against convergence to singularities and its speed of convergence is high. The deterministic annealing technique [134] also has a potential for overcoming the initialization problem of mixture models. Generally, log-likelihood functions to be maximized have many fixed points and multiple initialization approach is often used for avoiding local optimal solutions. On the other hand, a very fuzzy partition, in which all data samples belong to all clusters in some degree, is not so sensitive to initial partitioning. So, the FCM-type clustering algorithms have an advantage over mixture models when the graded tuning of the degree of fuzziness is used for estimating robust classification systems.

7.2 Probabilistic PCA Mixture Models and Regularized Fuzzy Clustering

7.2.1 Probabilistic Models for Principal Component Analysis

A useful tool for finding local features of large scale databases is local principal component analysis (local PCA) whose goal is to partition a data set into several small subregions and find linear expressions of the data subsets. Fukunaga and Olsen [37] proposed a two stage algorithm for local Karhunen-Loéve (KL) expansions including a clustering stage followed by KL expansion in each subregions. In the clustering stage, data partitioning is performed by using the clustering criterion of similarities among data points. Kambhatla and Leen [76] proposed an iterative algorithm composed of (hard) clustering of data sets and estimation of local principal components in each cluster.

Hinton *et al.* [52] extended the idea to "soft version", in which the responsibility of each data point for its generation is shared amongst all of the principal component analyzers instead of being assigned to only one analyzer. Although their PCA model is not exactly a probabilistic model defined by a certain probability density, the objective function to be minimized is given by a negative pseudo-likelihood function

$$J = \frac{1}{\nu}\sum_{i=1}^{c}\sum_{k=1}^{N}\psi_{ik}E(x_k, P_i) + \sum_{i=1}^{c}\sum_{k=1}^{N}\psi_{ik}\log\psi_{ik}, \tag{7.19}$$

where $E(x_k, P_i)$ is the squared reconstruction error, i.e., the distance between data point k and principal subspace P_i. In the local PCA model, the error criterion is used not only for estimating local principal components but also for estimating responsibilities of data points. The responsibility of the i-th analyzer for reconstructing data point x_k is given by

$$\psi_{ik} = \frac{\exp\left(-\frac{1}{\nu}E(x_k, P_i)\right)}{\sum_{j=1}^{c}\exp\left(-\frac{1}{\nu}E(x_k, P_j)\right)}. \tag{7.20}$$

In this way, local linear model estimation in conjunction with data partitioning is performed by minimizing a single negative likelihood function.

Roweis [136] and Tipping and Bishop [156] proposed probabilistic models for PCA and the single PCA model were extended to mixtures of local PCA models in which all of the model parameters are estimated through maximization of a single likelihood function. Mixture of probabilistic PCA (MPCA) [156] defines the linear latent models where a p dimensional observation vector x is related to a q dimensional latent vector f_i in each probabilistic model,

$$x = \mathsf{A}_i f_i + \mu_i + \epsilon_i; \qquad i = 1, \dots, c. \tag{7.21}$$

The $p \times q$ matrix A_i is the principal component matrix composed of q local principal component vectors and vector μ_i is the mean vector of the i-th probabilistic model. The density distribution of the latent variables is assumed to be a simple Gaussian that has zero-mean and unit variance, $f_i \sim \mathcal{N}(0, I_q)$ where I_q is $q \times q$ unit matrix. If the error model $\epsilon_i \sim \mathcal{N}(0, R_i)$ is associated to $R_i = \sigma_i^2 I_p$, the conventional PCA for the i-th local subspace is recovered with $\sigma_i \to 0$.

When we use isotropic Gaussian noise model of $\epsilon_i \sim \mathcal{N}(0, \sigma_i^2 I_p)$, the f_i conditional probability distribution over x space is given by $p_i(x|f_i) \sim \mathcal{N}(\mathsf{A}_i f_i + \mu_i, \sigma_i^2 I_p)$. The marginal distribution for observation x is also Gaussian:

$$p_i(x) \sim \mathcal{N}(\mu_i, W_i), \tag{7.22}$$

where the model covariance is $W_i = \mathsf{A}_i \mathsf{A}_i^\top + \sigma_i^2 I_p$. The log-likelihood function to be maximized is defined as

$$L(\Phi) = \sum_{k=1}^{N} \log\left\{\sum_{i=1}^{c} \alpha_i p_i(x_k)\right\}, \tag{7.23}$$

where $\Phi = (\alpha_1, \ldots, \alpha_c, \mu_1, \ldots, \mu_c, W_1, \ldots, W_c)$. The EM algorithm maximizes $L(\Phi)$ by maximizing the conditional expectation of the complete-data log-likelihood given a previous current estimate $\hat{\Phi}$ and $x_k, k = 1, \ldots, N$:

$$Q(\Phi|\hat{\Phi}) = \sum_{i=1}^{c} \sum_{k=1}^{N} \psi_{ik} \left\{\log \alpha_i + \log p_i(x_k)\right\}. \tag{7.24}$$

The parameters of these linear models are estimated by the EM algorithm.

The EM algorithm for MPCA

(O) Set an initial value $\Phi^{(0)}$ for the estimate. Let $\ell = 0$. Repeat the following **(E)** and **(M)** until convergence.

(E) (Expectation)
Calculate $Q(\Phi|\Phi^{(\ell)})$ by estimating the responsibility (posterior probability) of each data sample for component densities:

$$\psi_{ik} = \frac{\alpha_i \exp\left(-\frac{1}{2} E(x_k, \mu_i; W_i)\right) |W_i|^{-\frac{1}{2}}}{\sum_{j=1}^{c} \alpha_j \exp\left(-\frac{1}{2} E(x_k, \mu_j; W_j)\right) |W_j|^{-\frac{1}{2}}}, \tag{7.25}$$

where

$$E(x_k, \mu_i; W_i) = (x_k - \mu_i)^\top W_i^{-1} (x_k - \mu_i), \tag{7.26}$$

$$W_i = \mathsf{A}_i \mathsf{A}_i^\top + \sigma_i^2 I_p. \tag{7.27}$$

(M) (Maximization)
Find the maximizing solution

$$\bar{\Phi} = \arg\max_{\Phi} Q(\Phi|\Phi^{(\ell)})$$

by updating the parameters of components:

$$\alpha_i = \frac{1}{N}\sum_{k=1}^{N}\psi_{ik}, \tag{7.28}$$

$$\mu_i = \frac{\sum_{k=1}^{N}\psi_{ik}x_k}{\sum_{k=1}^{N}\psi_{ik}}, \tag{7.29}$$

$$\mathsf{A}_i = U_{qi}(\Delta_{qi} - \sigma_i^2 I_q)^{1/2}V, \tag{7.30}$$

$$\sigma_i^2 = \frac{1}{p-q}\sum_{j=q+1}^{p}\delta_{ij}, \tag{7.31}$$

where U_{qi} is the $p \times q$ matrix composed of the eigenvectors corresponding to the largest eigenvalues of the local responsibility-weighted covariance matrix S_i,

$$S_i = \frac{\sum_{k=1}^{N}\psi_{ik}(x_k - \mu_i)(x_k - \mu_i)^{\top}}{\sum_{k=1}^{N}\psi_{ik}}. \tag{7.32}$$

Δ_{qi} is the $q \times q$ diagonal matrix of the q largest eigenvalues and $\delta_{i,q+1}, \ldots, \delta_{ip}$ are the smallest $(p-q)$ eigenvalues of S_i. V is an arbitrary $q \times q$ orthogonal matrix.
Let $\ell := \ell + 1$ and $\Phi^{(\ell)} = \bar{\Phi}$.

The MPCA model is regarded as a constrained model of GMMs that captures the covariance structure of the p dimensional observation using $\mathsf{A}_i\mathsf{A}_i^{\top} + \sigma_i^2 I_p$ that has only $(p \times q + 1)$ free parameters while the full covariance matrix used in GMMs has p^2 parameters. The dimension of latent space q tunes the model complexity and the mixtures of latent variable models outperformed GMMs in terms of generalization.

7.2.2 Linear Fuzzy Clustering with Regularized Objective Function

When we use linear varieties as the prototypes of clusters in FCM-type fuzzy clustering, we have a linear fuzzy clustering algorithm called fuzzy c-varieties

(FCV) [7] that captures local linear structures of data sets. In the **FCV** algorithm, clustering criterion is the distance between data point k and the i-th linear variety P_i as follows:

$$D(x_k, P_i) = ||x_k - v_i||^2 - \sum_{\ell=1}^{q} |a_{i\ell}^\top (x_k - v_i)|^2, \tag{7.33}$$

where the q dimensional linear variety P_i spanned by basis vectors $a_{i\ell}$ is the prototype of cluster i. The optimal $a_{i\ell}$ are the eigenvectors corresponding to the largest eigenvalues of fuzzy scatter matrices. Thus, they are also regarded as the local principal component vectors. In the case of the entropy based method, the objective function is

$$J_{\text{efcv}}(U, P) = \sum_{i=1}^{c} \sum_{k=1}^{N} u_{ki} D(x_k, P_i) + \nu \sum_{i=1}^{c} \sum_{k=1}^{N} u_{ki} \log u_{ki}, \tag{7.34}$$

and is equivalent to the negative log-likelihood of Hinton's "soft version" local PCA.

Fuzzy c-elliptotypes (FCE) [7] is a hybrid of **FCM** and **FCV** where the clustering criterion is the linear mixture of two criteria:

$$\begin{aligned} D(x_k, P_i) &= (1-\alpha)||x_k - v_i||^2 + \alpha \left(||x_k - v_i||^2 - \sum_{\ell=1}^{q} |a_{i\ell}^\top (x_k - v_i)|^2 \right) \\ &= ||x_k - v_i||^2 - \alpha \sum_{\ell=1}^{q} |a_{i\ell}^\top (x_k - v_i)|^2. \end{aligned} \tag{7.35}$$

α is a predefined trade-off parameter that can vary from 0 for spherical shape clusters (**FCM**) to 1 for linear clusters (**FCV**).

In adaptive fuzzy c-elliptotypes (AFC) clustering [21], the trade-off parameter is also a decision variable so that the cluster shapes are tuned adaptively. The tuning, however, is not performed by minimization of a single objective function, i.e., the value of the objective function does not have the monotonically decreasing property in the adaptive method [160].

We can define the fuzzy counterpart of MPCA by applying the KL information based method to FCV clustering [55]. Replacing full rank matrix S_i with constrained matrix $W_i = \mathsf{A}_i \mathsf{A}_i^\top + \sigma_i^2 I_p$, the objective function of the **FCM** algorithm with KL information based method is modified as

$$\begin{aligned} J_{\text{klfcv}}(U, V, A, W) &= \sum_{i=1}^{c} \sum_{k=1}^{N} u_{ki} D(x_k, v_i; W_i) + \nu \sum_{i=1}^{c} \sum_{k=1}^{N} u_{ki} \log \frac{u_{ki}}{\alpha_i} \\ &\quad + \sum_{i=1}^{c} \sum_{k=1}^{N} u_{ki} \log |W_i|, \end{aligned} \tag{7.36}$$

where W is the total set of W_i. $D(x_k, v_i; W_i)$ is the generalized Mahalanobis distance

$$D(x_k, v_i; W_i) = (x_k - v_i)^\top W_i^{-1} (x_k - v_i). \tag{7.37}$$

Algorithm KLFCV: *FCV with KL information based method* [55]

KLFCV1. [Generate initial value:] Generate initial values for $\bar{V} = (\bar{v}_1, \dots, \bar{v}_c)$, $\bar{A} = (\bar{\alpha}_1, \dots, \bar{\alpha}_c)$ and $\bar{W} = (\bar{W}_1, \dots, \bar{W}_c)$.

KLFCV2. [Find optimal U:] Calculate

$$\bar{U} = \arg\min_{U \in \mathcal{U}_f} J_{\mathrm{klfcv}}(U, \bar{V}, \bar{A}, \bar{W}). \tag{7.38}$$

KLFCV3. [Find optimal V:] Calculate

$$\bar{V} = \arg\min_{V} J_{\mathrm{klfcv}}(\bar{U}, V, \bar{A}, \bar{W}). \tag{7.39}$$

KLFCV4. [Find optimal A:] Calculate

$$\bar{A} = \arg\min_{A \in \mathcal{A}} J_{\mathrm{klfcv}}(\bar{U}, \bar{V}, A, \bar{W}). \tag{7.40}$$

KLFCV5. [Find optimal W:] Calculate

$$\bar{W} = \arg\min_{W} J_{\mathrm{klfcv}}(\bar{U}, \bar{V}, \bar{A}, W). \tag{7.41}$$

KLFCV6. [Test convergence:] If all parameters are convergent, stop; else go to KLFCV2.

End KLFCV.

It is easy to see that new memberships are derived as

$$u_{ki} = \frac{\alpha_i \exp\left(-\frac{1}{\nu} D(x_k, v_i; W_i)\right) |W_i|^{-\frac{1}{\nu}}}{\sum_{j=1}^{c} \alpha_j \exp\left(-\frac{1}{\nu} D(x_k, v_j; W_j)\right) |W_j|^{-\frac{1}{\nu}}}, \tag{7.42}$$

and v_i and α_i are updated by using Eqs.(7.16) and (7.17), respectively.

The solution of the optimal W is as follows:

$$W_i = \mathsf{A}_i \mathsf{A}_i^\top + \sigma_i^2 I_p \tag{7.43}$$

$$\mathsf{A}_i = U_{qi}(\Delta_{qi} - \sigma_i^2 I_q)^{1/2} V \tag{7.44}$$

$$\sigma_i^2 = \frac{1}{p-q} \sum_{j=q+1}^{p} \delta_{ij} \tag{7.45}$$

Proof. To calculate new A_i and σ_i, we rewrite the objective function as

$$\begin{aligned} J_{\mathrm{klfcv}'}(U, V, A, W) = & \sum_{i=1}^{c} \left\{ \sum_{k=1}^{N} u_{ki} \right\} \mathrm{tr}(W_i^{-1} S_i) \\ & + \nu \sum_{i=1}^{c} \sum_{k=1}^{N} u_{ki} \log \frac{u_{ki}}{\alpha_i} \\ & + \sum_{i=1}^{c} \left\{ \sum_{k=1}^{N} u_{ki} \right\} \log |W_i|, \end{aligned} \tag{7.46}$$

where S_i is the fuzzy covariance matrix in cluster i that is calculated by (7.18). Let denote zero-matrix as O. From the necessary condition $\partial J_{\mathrm{klfcv'}}/\partial \mathsf{A}_i = O$,

$$-W_i^{-1}S_iW_i^{-1}\mathsf{A}_i + W_i^{-1}\mathsf{A}_i = O. \tag{7.47}$$

Equation (7.47) is the same as the necessary condition for the MPCA model. Then, the local principal component matrix A_i is derived as follows:

$$\mathsf{A}_i = U_{qi}(\Delta_{qi} - \sigma_i^2 I_q)^{1/2}V. \tag{7.48}$$

This is the same equation as (7.30) and the optimal A_i are given by the eigenvectors corresponding to the largest eigenvalues. The optimal σ_i^2 is also derived as

$$\sigma_i^2 = \frac{1}{p-q}\sum_{j=q+1}^{p} \delta_{ij}. \tag{7.49}$$

Thus, (7.44) and (7.45) are obtained. (For more details, see [55, 156]). □

While this constrained **FCM** algorithm is equivalent to the MPCA algorithm in the case of $\nu = 2$, there is no corresponding probabilistic models when $\nu \neq 2$. Then the method is not a probabilistic approach but a sort of fuzzy modeling technique where the parameter ν determines the degree of fuzziness. This algorithm is regarded as a version of the AFC clustering, where the additional parameters α_i and σ_i, $i = 1, \ldots, c$, play a role for tuning the cluster shape adaptively and minimizing a single objective function achieves the optimization of the parameters.

7.2.3 An Illustrative Example

Let us demonstrate the characteristic features of the KL information based method in a simple illustrative example [55]. The artificial data shown in Fig. 7.1 includes two data sub-groups and two linear generative models should be identified. One is the linear model whose error variance is small and the data points are distributed forming a thin line. The other set is generated with larger error model and the data points form a rectangle. We compare the classification functions derived by MPCA and **FCV** with the KL information based method, in which the dimensionality of the latent variable is 1.

Fig. 7.2 shows the result of MPCA that is equivalent to the KLFCV model with $\nu = 2$. The vertical axis shows the value of classification function (membership degree) for 1st cluster. The data set was classified into two clusters represented by circles and crosses in the sense of maximum membership (posterior probability), and the cluster volumes (a priori probabilities) were $\alpha_1 = 0.1$ and $\alpha_2 = 0.9$, respectively. Because $\alpha_1 << \alpha_2$, the probabilistic model regarded the 2nd cluster as a meaningful cluster and classified the overlapped region as the 2nd cluster.

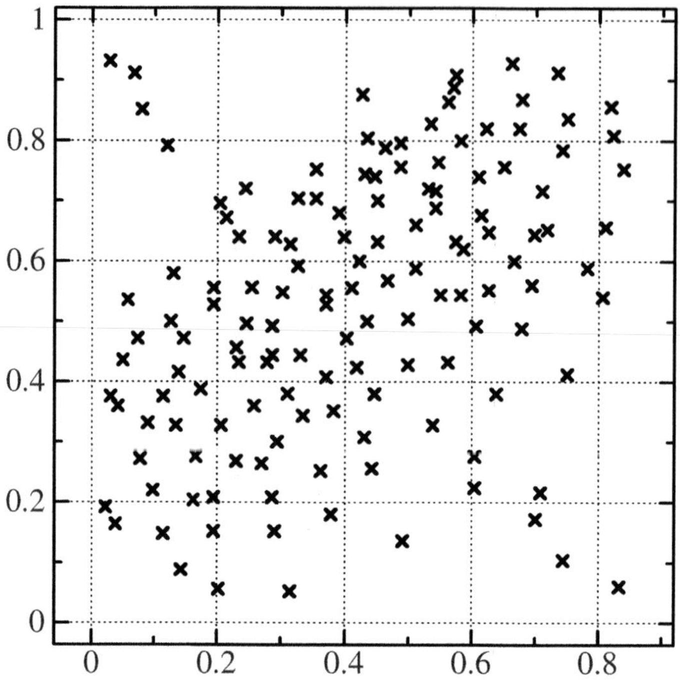

Fig. 7.1. An example of two linear generative models

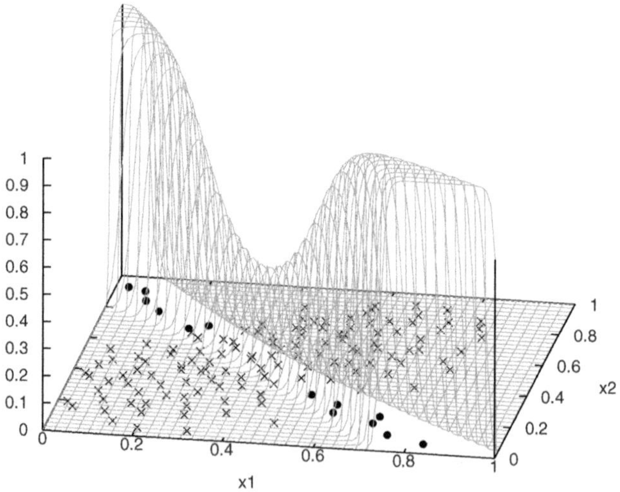

Fig. 7.2. Result by MPCA (**KLFCV** with $\nu = 2$)

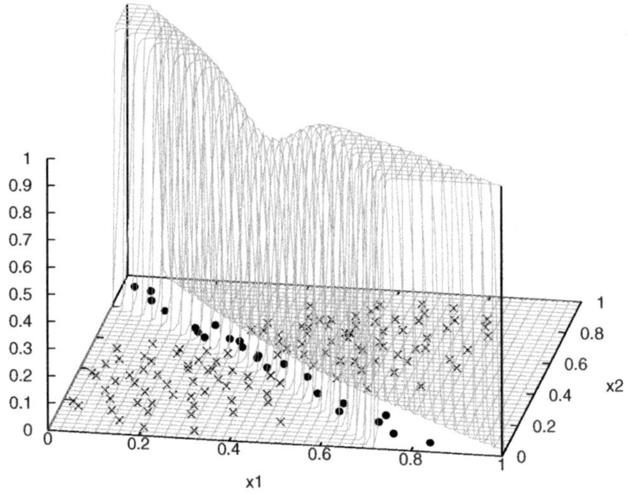

Fig. 7.3. Result by **KLFCV** with $\nu = 1$

Fig. 7.3 shows the result of **KLFCV** with $\nu = 1$ where $\alpha_1 = 0.15$ and $\alpha_2 = 0.85$, respectively. Because $\nu < 2$, the derived partitioning was not so fuzzy as that of MPCA and tended toward the crisp one emphasizing the smaller cluster. From the classification view point, a crisp model is more useful than fuzzy model and the classification function is suited for our intuitive partitioning. In this way, the fuzzy clustering model can derive more flexible partitioning by tuning the parameter ν.

8 Local Multivariate Analysis Based on Fuzzy Clustering

In this chapter, we describe the close relationship between enhanced clustering algorithms and local multivariate analysis. The objective functions are defined based on the standard fuzzification approach. However, it is easy to define similar objective functions based on other fuzzification approaches and the most parts of the discussions given in this chapter also apply to them.

8.1 Switching Regression and Fuzzy c-Regression Models

Switching regression is a technique for applying regression analysis in conjunction with data partitioning, in which estimation of regression models in local areas is performed after preprocessing of stratified sampling (classification of samples). Fuzzy c-regression models (FCRM) [48] performs switching regression and FCM-type clustering simultaneously, and the measure of error in regression models is used not only for estimating parameters of regression models but also for estimating fuzzy c-partition in FCM-type fuzzy clustering.

In this section, we introduce two different formulations of **FCRM**.

8.1.1 Linear Regression Model

Let $(x_k, y_k), k = 1, \dots, N$ be N data samples where each p-dimensional independent observation x_k is given with its corresponding r-dimensional dependent observation y_k. The goal of linear regression analysis is to reveal the correlation structure between independent observation and dependent observation.

Linear regression model including y-intercept. First, we consider the following linear regression model:

$$
\begin{aligned}
y_k^1 &= \beta_1^0 + \beta_1^1 x_k^1 + \cdots + \beta_1^p x_k^p + \varepsilon^1 \\
y_k^2 &= \beta_2^0 + \beta_2^1 x_k^1 + \cdots + \beta_2^p x_k^p + \varepsilon^2 \\
&\cdots\cdots\cdots \\
y_k^r &= \beta_r^0 + \beta_r^1 x_k^1 + \cdots + \beta_r^p x_k^p + \varepsilon^r
\end{aligned}
$$

S. Miyamoto et al.: Algorithms for Fuzzy Clustering, STUDFUZZ 229, pp. 171–194, 2008.
springerlink.com

Here, $\beta_{\cdot} = (\beta_{\cdot}^0, \beta_{\cdot}^1, \ldots, \beta_{\cdot}^p)^\top$ are the "partial regression coefficient vector" to be estimated, and $\varepsilon = (\varepsilon^1, \ldots, \varepsilon^r)$ is an isotropic Gaussian observation error. Based on the least squares method, we estimate the model parameters so that the sum of squared regression errors J_{lra1} is minimized.

$$J_{\mathrm{lra1}}(B_0) = \sum_{j=1}^{r} \sum_{k=1}^{N} (y_k^j - \beta_j^0 - \beta_j^1 x_k^1 - \cdots - \beta_j^p x_k^p)^2, \tag{8.1}$$

where $B_0 = (\beta_1, \ldots, \beta_r)$ is the total set of partial regression coefficient vectors. Using matrix multiplication, we can write the regression model as

$$\mathsf{Y} = \mathsf{Z}\mathsf{B}_0 + \mathsf{E}, \tag{8.2}$$

$$\mathsf{Y} = \begin{pmatrix} y_1^1 & y_1^2 & \cdots & y_1^r \\ y_2^1 & y_2^2 & \cdots & y_2^r \\ \vdots & \vdots & \ddots & \vdots \\ y_N^1 & y_N^2 & \cdots & y_N^r \end{pmatrix}, \quad \mathsf{Z} = \begin{pmatrix} 1 & x_1^1 & x_1^2 & \cdots & x_1^p \\ 1 & x_2^1 & x_2^2 & \cdots & x_2^p \\ \vdots & \vdots & \vdots & \ddots & \vdots \\ 1 & x_N^1 & x_N^2 & \cdots & x_N^p \end{pmatrix},$$

$$\mathsf{B}_0 = \begin{pmatrix} \beta_1^0 & \beta_2^0 & \cdots & \beta_r^0 \\ \beta_1^1 & \beta_2^1 & \cdots & \beta_r^1 \\ \beta_1^2 & \beta_2^2 & \cdots & \beta_r^2 \\ \vdots & \vdots & \ddots & \vdots \\ \beta_1^p & \beta_2^p & \cdots & \beta_r^p \end{pmatrix}, \quad \mathsf{E} = \begin{pmatrix} \varepsilon_1^1 & \varepsilon_1^2 & \cdots & \varepsilon_1^r \\ \varepsilon_2^1 & \varepsilon_2^2 & \cdots & \varepsilon_2^r \\ \vdots & \vdots & \ddots & \vdots \\ \varepsilon_N^1 & \varepsilon_N^2 & \cdots & \varepsilon_N^r \end{pmatrix}.$$

Let O be the zero-matrix whose all elements are zero. From the necessary condition

$$\frac{\partial J_{\mathrm{lra1}}}{\partial \mathsf{B}_0} = -2\mathsf{Z}^\top \mathsf{Y} + 2\mathsf{Z}^\top \mathsf{Z}\mathsf{B}_0 = O, \tag{8.3}$$

we have the following normal equation:

$$\mathsf{Z}^\top \mathsf{Z}\mathsf{B}_0 = \mathsf{Z}^\top \mathsf{Y}. \tag{8.4}$$

Under the assumption that $\mathsf{Z}^\top\mathsf{Z}$ has its inverse $(\mathsf{Z}^\top\mathsf{Z})^{-1}$, i.e., $\mathsf{Z}^\top\mathsf{Z}$ is a regular matrix, the optimal partial regression coefficient matrix is given as

$$\mathsf{B}_0 = (\mathsf{Z}^\top \mathsf{Z})^{-1}\mathsf{Z}^\top \mathsf{Y}. \tag{8.5}$$

Linear regression model with centered data. Second, we consider the linear regression model for normalized data where each observed element is subtracted by its corresponding mean value so that each observation has zero-mean [27, 81].

$$\begin{aligned} y_k^1 - \bar{y}^1 &= \beta_1^1(x_k^1 - \bar{x}^1) + \cdots + \beta_1^p(x_k^p - \bar{x}^p) + \varepsilon^1 \\ y_k^2 - \bar{y}^2 &= \beta_2^1(x_k^1 - \bar{x}^1) + \cdots + \beta_2^p(x_k^p - \bar{x}^p) + \varepsilon^2 \\ &\cdots\cdots\cdots \\ y_k^r - \bar{y}^r &= \beta_r^1(x_k^1 - \bar{x}^1) + \cdots + \beta_r^p(x_k^p - \bar{x}^p) + \varepsilon^r \end{aligned}$$

Here, $\bar{y} = (\bar{y}^1, \ldots, \bar{y}^r)$ and $\bar{x} = (\bar{x}^1, \ldots, \bar{x}^p)$ are the mean vector, and $\beta_\cdot = (\beta_\cdot^1, \ldots, \beta_\cdot^p)^\top$ are the "partial regression coefficient vector" to be estimated. The above equations do not include y-intercept, and the goal is to search for the best fitting straight line that passes through the origin. The sum of squared regression errors J_{lra2} is written as

$$J_{\mathrm{lra2}}(B) = \sum_{j=1}^{r} \sum_{k=1}^{N} \left(y_k^j - \bar{y}^j - \beta_j^1 (x_k^1 - \bar{x}^1) - \cdots - \beta_j^p (x_k^p - \bar{x}^p) \right)^2, \quad (8.6)$$

where $B = (\beta_1, \ldots, \beta_r)$ is the total set of partial regression coefficient vectors. Using matrix multiplication, we can write the regression model as

$$\hat{\mathsf{Y}} = \hat{\mathsf{X}}\mathsf{B} + \mathsf{E}, \quad (8.7)$$

$$\hat{\mathsf{Y}} = \begin{pmatrix} y_1^1 - \bar{y}^1 & y_1^2 - \bar{y}^2 & \ldots & y_1^r - \bar{y}^r \\ y_2^1 - \bar{y}^1 & y_2^2 - \bar{y}^2 & \ldots & y_2^r - \bar{y}^r \\ \vdots & \vdots & \ddots & \vdots \\ y_N^1 - \bar{y}^1 & y_N^2 - \bar{y}^2 & \ldots & y_N^r - \bar{y}^r \end{pmatrix},$$

$$\hat{\mathsf{X}} = \begin{pmatrix} x_1^1 - \bar{x}^1 & x_1^2 - \bar{x}^2 & \ldots & x_1^p - \bar{x}^p \\ x_2^1 - \bar{x}^1 & x_2^2 - \bar{x}^2 & \ldots & x_2^p - \bar{x}^p \\ \vdots & \vdots & \ddots & \vdots \\ x_N^1 - \bar{x}^1 & x_N^2 - \bar{x}^2 & \ldots & x_N^p - \bar{x}^p \end{pmatrix},$$

$$\mathsf{B} = \begin{pmatrix} \beta_1^1 & \beta_2^1 & \ldots & \beta_r^1 \\ \beta_1^2 & \beta_2^2 & \ldots & \beta_r^2 \\ \vdots & \vdots & \ddots & \vdots \\ \beta_1^p & \beta_2^p & \ldots & \beta_r^p \end{pmatrix}, \qquad \mathsf{E} = \begin{pmatrix} \varepsilon_1^1 & \varepsilon_1^2 & \ldots & \varepsilon_1^r \\ \varepsilon_2^1 & \varepsilon_2^2 & \ldots & \varepsilon_2^r \\ \vdots & \vdots & \ddots & \vdots \\ \varepsilon_N^1 & \varepsilon_N^2 & \ldots & \varepsilon_N^r \end{pmatrix}.$$

From the necessary condition

$$\frac{\partial J_{\mathrm{lra2}}}{\partial \mathsf{B}} = -2\hat{\mathsf{X}}^\top \hat{\mathsf{Y}} + 2\hat{\mathsf{X}}^\top \hat{\mathsf{X}}\mathsf{B} = O, \quad (8.8)$$

we have the following normal equation:

$$\hat{\mathsf{X}}^\top \hat{\mathsf{X}}\mathsf{B} = \hat{\mathsf{X}}^\top \hat{\mathsf{Y}}. \quad (8.9)$$

Under the assumption that $\hat{\mathsf{X}}^\top \hat{\mathsf{X}}$ has its inverse $(\hat{\mathsf{X}}^\top \hat{\mathsf{X}})^{-1}$, i.e., $\hat{\mathsf{X}}^\top \hat{\mathsf{X}}$ is a regular matrix, the optimal partial regression coefficient matrix is given as

$$\mathsf{B} = (\hat{\mathsf{X}}^\top \hat{\mathsf{X}})^{-1} \hat{\mathsf{X}}^\top \hat{\mathsf{Y}}. \quad (8.10)$$

Unified view of two regression models. Third, we show that the above two regression models are identical. From (8.4), we can rewrite the normal equation for the j-th dependent observation as

$$\beta_j^0 N + \beta_j^1 \sum_{k=1}^{N} x_k^1 + \cdots + \beta_j^p \sum_{k=1}^{N} x_k^p = \sum_{k=1}^{N} y_k^j$$

$$\beta_j^0 \sum_{k=1}^{N} x_k^1 + \beta_j^1 \sum_{k=1}^{N} (x_k^1)^2 + \cdots + \beta_j^p \sum_{k=1}^{N} x_k^1 x_k^p = \sum_{k=1}^{N} x_k^1 y_k^j$$

$$\cdots\cdots\cdots$$

$$\beta_j^0 \sum_{k=1}^{N} x_k^p + \beta_j^1 \sum_{k=1}^{N} x_k^1 x_k^p + \cdots + \beta_j^p \sum_{k=1}^{N} (x_k^p)^2 = \sum_{k=1}^{N} x_k^p y_k^j$$

and the first equation derives β_j^0 as

$$\begin{aligned} \beta_j^0 &= \frac{1}{N} \sum_{k=1}^{N} y_k^j - \beta_j^1 \frac{1}{N} \sum_{k=1}^{N} x_k^1 + \cdots + \beta_j^p \frac{1}{N} \sum_{k=1}^{N} x_k^p \\ &= \bar{y}^j - \beta_j^1 \bar{x}^1 - \cdots - \beta_j^p \bar{x}^p. \end{aligned} \tag{8.11}$$

Then, the remaining equations can be represented as

$$\beta_j^1 \left(\sum_{k=1}^{N} (x_k^1)^2 - N(\bar{x}^1)^2 \right) + \cdots + \beta_j^p \left(\sum_{k=1}^{N} x_k^1 x_k^p - N\bar{x}^1 \bar{x}^p \right) = \sum_{k=1}^{N} x_k^1 y_k^j - N\bar{x}^1 \bar{y}^j$$

$$\cdots\cdots\cdots$$

$$\beta_j^1 \left(\sum_{k=1}^{N} x_k^1 x_k^p - N\bar{x}^1 \bar{x}^p \right) + \cdots + \beta_j^p \left(\sum_{k=1}^{N} (x_k^p)^2 - N(\bar{x}^p)^2 \right) = \sum_{k=1}^{N} x_k^p y_k^j - N\bar{x}^p \bar{y}^j$$

It is obvious that these equations are equivalent to (8.9). Therefore, we can derive the same $\beta_j^1, \beta_j^2, \ldots, \beta_j^p$ using two regression models. Sequentially, the regression model including y-intercept can be transformed as

$$\begin{aligned} y_k^j &= \beta_j^0 + \beta_j^1 x_k^1 + \cdots + \beta_j^p x_k^p + \varepsilon_k^j \\ &= \bar{y}^j - \beta_j^1 \bar{x}^1 - \cdots - \beta_j^p \bar{x}^p + \beta_j^1 x_k^1 + \cdots + \beta_j^p x_k^p + \varepsilon_k^j \\ &= \bar{y}^j + \beta_j^1 (x_k^1 - \bar{x}^1) + \cdots + \beta_j^p (x_k^p - \bar{x}^p) + \varepsilon_k^j. \end{aligned} \tag{8.12}$$

In this way, the two regression models are identical. See [27, 81] for more detailed discussion.

8.1.2 Switching Linear Regression by Standard Fuzzy c-Regression Models

Here, we consider a switching linear regression model. In **FCRM**, it is assumed that the data set was drawn from c models:

$$\mathsf{Y} = \mathsf{Z}\mathsf{B}_{0i} + \mathsf{E}_i, \quad i = 1, \ldots, c, \tag{8.13}$$

where B_{0i} is the regression coefficient matrix in the i-th model and E_i is a corresponding error matrix. Then, the objective function to be minimized is

$$\begin{aligned} J_{\mathrm{fcrm1}}(U, B_0) &= \sum_{i=1}^{c}\sum_{k=1}^{N}(u_{ki})^m D((x_k, y_k), \mathsf{B}_{0i}) \\ &= \sum_{i=1}^{c}\sum_{k=1}^{N}(u_{ki})^m ||y_k^\top - z_k^\top \mathsf{B}_{0i}||^2 \\ &= \sum_{i=1}^{c} \mathrm{tr}\Big((\mathsf{Y} - \mathsf{Z}\mathsf{B}_{0i})^\top \mathsf{U}_i^m (\mathsf{Y} - \mathsf{Z}\mathsf{B}_{0i})\Big), \end{aligned} \tag{8.14}$$

where "tr" represents the trace of matrices (the sum of diagonal entries). U and B_0 are the total sets of memberships u_{ki} and regression coefficient matrices B_{0i}, respectively. $D((x_k, y_k), \mathsf{B}_{0i})$ is the measure of error in the i-th model and $z_k^\top$ is the k-th row of Z. U_i is the $N \times N$ diagonal matrix whose k-th diagonal element is u_{ki}. The goal is to estimate the regression model parameters B_{0i} and the fuzzy memberships u_{ki}. Considering the necessary condition for the optimality, the normal equation is given as

$$(\mathsf{Z}^\top \mathsf{U}_i^m \mathsf{Z})\mathsf{B}_{0i} = \mathsf{Z}^\top \mathsf{U}_i^m \mathsf{Y}. \tag{8.15}$$

Then, the updating rule for B_{0i} is derived as

$$\mathsf{B}_{0i} = (\mathsf{Z}^\top \mathsf{U}_i^m \mathsf{Z})^{-1}\mathsf{Z}^\top \mathsf{U}_i^m \mathsf{Y}. \tag{8.16}$$

In the same way, new membership u_{ki} is given by

$$u_{ki} = \left[\sum_{j=1}^{c}\left(\frac{D((x_k, y_k), \mathsf{B}_{0i})}{D((x_k, y_k), \mathsf{B}_{0j})}\right)^{\frac{1}{m-1}}\right]^{-1}, \tag{8.17}$$

where the clustering criterion of **FCM** is replaced with regression error $D((x_k, y_k), \mathsf{B}_{0i})$.

8.1.3 Local Regression Analysis with Centered Data Model

We can also define another formulation for switching regression considering normalization in each cluster as follows:

$$\begin{aligned} y_k^1 - v_{yi}^1 &= \beta_{i1}^1(x_k^1 - v_{xi}^1) + \cdots + \beta_{i1}^p(x_k^p - v_{xi}^p) + \varepsilon_i^1 \\ y_k^2 - v_{yi}^2 &= \beta_{i2}^1(x_k^1 - v_{xi}^1) + \cdots + \beta_{i2}^p(x_k^p - v_{xi}^p) + \varepsilon_i^2 \\ &\cdots\cdots\cdots \\ y_k^r - v_{yi}^r &= \beta_{ir}^1(x_k^1 - v_{xi}^1) + \cdots + \beta_{ir}^p(x_k^p - v_{xi}^p) + \varepsilon_i^r \end{aligned}$$

where $v_{xi} = (v_{xi}^1, \ldots, v_{xi}^p)^\top$ and $v_{yi} = (v_{yi}^1, \ldots, v_{yi}^r)^\top$ are the center (mean vector) of cluster i. Using matrix multiplication, we can write the regression model as

$$\mathsf{Y} - 1_N v_{yi}^\top = (\mathsf{X} - 1_N v_{xi}^\top)\mathsf{B}_i + \mathsf{E}_i, \tag{8.18}$$

where B_i is the regression coefficient matrix in cluster i. 1_N is the N-dimensional vector whose all elements are 1. Then, the objective function of **FCRM** is also defined as

$$\begin{aligned} J_{\mathrm{fcrm2}}(U, V, B) &= \sum_{i=1}^{c}\sum_{k=1}^{N}(u_{ki})^m D((x_k, y_k), (v_{xi}, v_{yi}); \mathsf{B}_i) \\ &= \sum_{i=1}^{c}\sum_{k=1}^{N}(u_{ki})^m \|y_k^\top - v_{yi}^\top - (x_k - v_{xi})^\top \mathsf{B}_i\|^2 \\ &= \sum_{i=1}^{c} \mathrm{tr}\Big((\mathsf{Y}_i - \mathsf{X}_i\mathsf{B}_i)^\top \mathsf{U}_i^m (\mathsf{Y}_i - \mathsf{X}_i\mathsf{B}_i)\Big), \end{aligned} \tag{8.19}$$

where V and B are the total sets of cluster centers (v_{xi}, v_{yi}) and regression coefficient matrices B_i, respectively. $\mathsf{Y}_i = \mathsf{Y} - 1_N v_{yi}^\top$ and $\mathsf{X}_i = \mathsf{X} - 1_N v_{xi}^\top$.

Necessary condition for the optimality $\partial J_{\mathrm{fcrm2}}/\partial \mathsf{B}_i = O$ reduces the normal equation as

$$(\mathsf{X}_i^\top \mathsf{U}_i^m \mathsf{X}_i)\mathsf{B}_i = \mathsf{X}_i^\top \mathsf{U}_i^m \mathsf{Y}_i, \tag{8.20}$$

and the optimal B_i is given as

$$\mathsf{B}_i = (\mathsf{X}_i^\top \mathsf{U}_i^m \mathsf{X}_i)^{-1} \mathsf{X}_i^\top \mathsf{U}_i^m \mathsf{Y}_i. \tag{8.21}$$

In the same way, the membership is updated in a same equation with (8.17) using $D((x_k, y_k), (v_{xi}, v_{yi}); \mathsf{B}_i)$ instead of $D((x_k, y_k), \mathsf{B}_{0i})$.

Let 0_p be p dimensional zero vector. Next, the conditions $\partial J_{\mathrm{fcrm2}}/\partial v_{xi} = 0_p$ and $\partial J_{\mathrm{fcrm2}}/\partial v_{yi} = 0_r$ derives

$$\frac{\sum_{k=1}^{N}(u_{ki})^m y_k^\top}{\sum_{k=1}^{N}(u_{ki})^m} - \frac{\sum_{k=1}^{N}(u_{ki})^m x_k^\top \mathsf{B}_i}{\sum_{k=1}^{N}(u_{ki})^m} = v_{yi}^\top - v_{xi}^\top \mathsf{B}_i, \tag{8.22}$$

and the equation means that the optimal v_{xi} and v_{yi} can be arbitrary selected on the regression surface that passes through the cluster center,

$$\left(\frac{\sum_{k=1}^{N}(u_{ki})^m x_k}{\sum_{k=1}^{N}(u_{ki})^m}, \frac{\sum_{k=1}^{N}(u_{ki})^m y_k}{\sum_{k=1}^{N}(u_{ki})^m} \right). \tag{8.23}$$

Then, a natural choice is

$$v_{xi} = \frac{\sum_{k=1}^{N}(u_{ki})^m x_k}{\sum_{k=1}^{N}(u_{ki})^m} \tag{8.24}$$

and

$$v_{yi} = \frac{\sum_{k=1}^{N} (u_{ki})^m y_k}{\sum_{k=1}^{N} (u_{ki})^m}. \tag{8.25}$$

Note that the updating formulas are equivalent to that of **FCM**.

8.1.4 Connection of the Two Formulations

We can also show that the above two switching regression models are identical. In the standard FCRM model, we can rewrite the normal equation for the j-th dependent observation in cluster i as

$$\beta_{ij}^0 \sum_{k=1}^{N} (u_{ki})^m + \beta_{ij}^1 \sum_{k=1}^{N} (u_{ki})^m x_k^1 + \cdots + \beta_{ij}^p \sum_{k=1}^{N} (u_{ki})^m x_k^p = \sum_{k=1}^{N} (u_{ki})^m y_k^j$$

$$\beta_{ij}^0 \sum_{k=1}^{N} (u_{ki})^m x_k^1 + \beta_{ij}^1 \sum_{k=1}^{N} (u_{ki})^m (x_k^1)^2 + \cdots + \beta_{ij}^p \sum_{k=1}^{N} (u_{ki})^m x_k^1 x_k^p = \sum_{k=1}^{N} (u_{ki})^m x_k^1 y_k^j$$

$$\cdots\cdots\cdots$$

$$\beta_{ij}^0 \sum_{k=1}^{N} (u_{ki})^m x_k^p + \beta_{ij}^1 \sum_{k=1}^{N} (u_{ki})^m x_k^1 x_k^p + \cdots + \beta_{ij}^p \sum_{k=1}^{N} (u_{ki})^m (x_k^p)^2 = \sum_{k=1}^{N} (u_{ki})^m x_k^p y_k^j$$

and the first equation derives β_{ij}^0 as

$$\beta_{ij}^0 = \frac{\sum_{k=1}^{N} (u_{ki})^m y_k^j}{\sum_{k=1}^{N} (u_{ki})^m} - \beta_{ij}^1 \frac{\sum_{k=1}^{N} (u_{ki})^m x_k^1}{\sum_{k=1}^{N} (u_{ki})^m} - \cdots - \beta_{ij}^p \frac{\sum_{k=1}^{N} (u_{ki})^m x_k^p}{\sum_{k=1}^{N} (u_{ki})^m}$$
$$= v_{yi}^j - \beta_{ij}^1 v_{xi}^1 - \cdots - \beta_{ij}^p v_{xi}^p. \tag{8.26}$$

Then, the remaining equations can be represented as

$$(\mathsf{X}_i^\top \mathsf{U}_i^m \mathsf{X}_i)\mathsf{B}_i = \mathsf{X}_i^\top \mathsf{U}_i^m \mathsf{Y}_i. \tag{8.27}$$

Therefore, we can derive the same $\beta_{ij}^1, \beta_{ij}^2, \ldots, \beta_{ij}^p$ using two regression models. In this way, the two regression models are identical.

8.1.5 An Illustrative Example

Let us compare the switching regression lines derived by two FCRM formulations. The artificial data shown in Fig. 8.1 includes two data sub-groups and two regression lines should be identified.

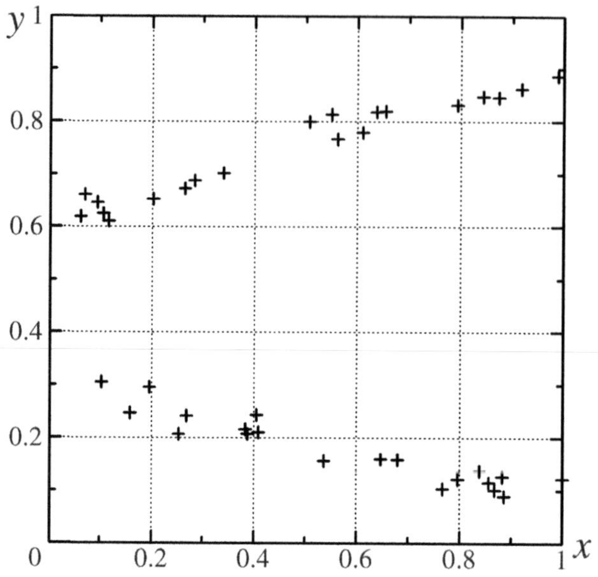

Fig. 8.1. An example of two linear generative model

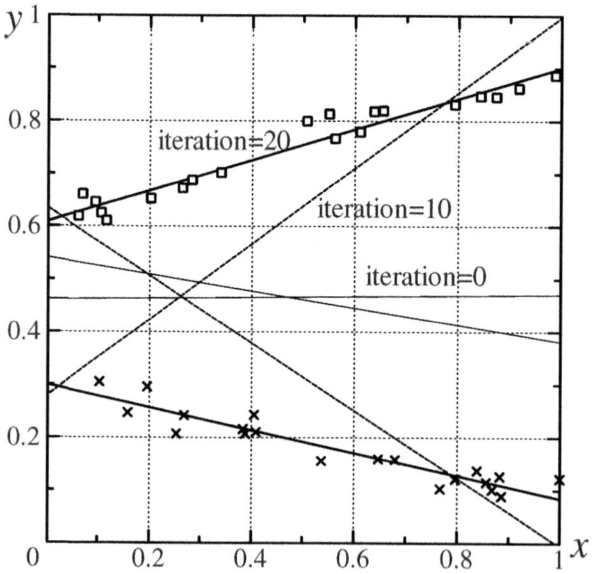

Fig. 8.2. Result by standard **FCRM**

Figures 8.2 and 8.3 show the switching regression lines derived from same initial partition by the standard FCRM formulation and the FCRM formulation based on the centered data model, respectively. The clustering parameters were

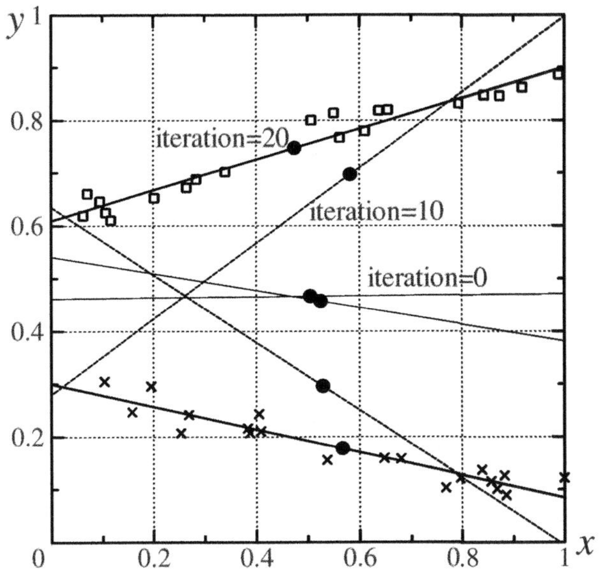

Fig. 8.3. Result by centered data **FCRM**

set as $c = 2$ and $m = 2$. In Fig. 8.3, black circles depict the cluster center of each cluster. Because the two formulations are identical, the clustering results in intermediate steps are also same. Furthermore, Fig. 8.3 shows that cluster centers in **FCRM** are useful for summarizing data sub-structures as well.

8.2 Local Principal Component Analysis and Fuzzy c-Varieties

Fuzzy c-varieties (FCV) proposed by Bezdek *et al.* [6, 7] is a linear fuzzy clustering algorithm that partitions a data set into several linear clusters using linear varieties as prototypes of clusters. Although the goal of the FCV clustering is unsupervised classification of non-labeled data sets, it can be regarded as a simultaneous approach to principal component analysis (PCA) and fuzzy clustering since the basis vectors of prototypical linear varieties are often identified with local principal component vectors [55].

For performing PCA, several formulations have been proposed. This section considers four major formulations, in which principal component vectors correspond to the eigenvectors of covariance matrix, and discusses close relation between local PCA and FCV clustering.

8.2.1 Several Formulations for Principal Component Analysis

Let $x_k = (x_k^1, \ldots, x_k^p)^\top$, $k = 1, \ldots, N$ denote p dimensional observation of N samples, and $f_k = (f_k^1, \ldots, f_k^q)^\top$, $k = 1, \ldots, N$ be q ($q < p$) dimensional latent variables to be estimated. The data vectors can also be represented as $N \times p$ data matrix $\mathsf{X} = (x_1, \ldots, x_N)^\top$.

Fitting low-dimensional subspace. Pearson [128] defined the PCA problem as the task of finding the "best fitting" straight line or plane that minimizes the sum of distances between data points and the line or plane. So, the objective function to be minimized is given as follows:

$$J_{\text{pca1}}(A) = \sum_{k=1}^{N} \left(||x_k - v||^2 - \sum_{\ell=1}^{q} |a_\ell^\top (x_k - v)|^2 \right), \tag{8.28}$$

where v is the mean vector and $A = (a_1, \ldots, a_q)$ are the set of the orthogonal basis vector of the linear variety to be found. The optimal a_ℓ having unit length is derived by Lagrange multiplier method, and the problem is reduced to the following eigenvalue problem:

$$S a_\ell = \gamma_\ell a_\ell, \tag{8.29}$$

where S is the scatter matrix (a constant multiple of the covariance matrix)

$$S = \sum_{k=1}^{N} (x_k - v)(x_k - v)^\top. \tag{8.30}$$

Then, under the constraint of $||a_\ell|| = 1$, (8.28) is rewritten as

$$\begin{aligned} J_{\text{pca1}}(A) &= \sum_{k=1}^{N} ||x_k - v||^2 - \sum_{\ell=1}^{q} a_\ell^\top S a_\ell \\ &= \sum_{k=1}^{N} ||x_k - v||^2 - \sum_{\ell=1}^{q} \gamma_\ell. \end{aligned} \tag{8.31}$$

Here, we should choose the q maximum eigenvalues and their corresponding eigenvectors because the first term is constant with fixed v.

Maximization of variance of latent variable. If we wish to estimate latent variables so as to keep the original information as well as possible, we should estimate the latent variables that have the largest variances. So, in the PCA problem, minimization of the distances between data points and low-dimensional subspace is often performed by maximizing the variance of latent variables because the first term of (8.31) is invariant with the fixed mean vector.

Lower rank approximation of data matrix. Another formulation for PCA is defined by using the least squares criterion [167]. This formulation can be regarded as a constrained model of factor analysis. In order to perform dimension reduction with the minimum loss of information, the least squares criterion to be minimized is defined as

$$J_{\text{pca3}}(\mathsf{F}, \mathsf{A}) = \text{tr}\left\{ (\mathsf{X} - \mathsf{Y})^\top (\mathsf{X} - \mathsf{Y}) \right\}, \tag{8.32}$$

where "tr" represents the trace of matrices (the sum of diagonal entries). $\mathsf{Y} = (y_1, \ldots, y_N)^\top$ denotes the lower rank approximation of the data matrix X,

$$\mathsf{Y} = \mathsf{F}\mathsf{A}^\top + 1_N v^\top, \tag{8.33}$$

where $\mathsf{F} = (f_1, \ldots, f_N)^\top$ is the $N \times q$ score matrix and $\mathsf{A} = (a_1, \ldots, a_q)$ is the $p \times q$ principal component matrix. 1_N is the N dimensional vector whose all elements are 1. Let I_q be $q \times q$ unit matrix and 0_q be q dimensional zero vector. Usually, the parameters are estimated under the following conditions:

$$\mathsf{F}^\top \mathsf{F} = I_q, \tag{8.34}$$

$$\mathsf{F}^\top 1_N = 0_q, \tag{8.35}$$

and $\mathsf{A}^\top \mathsf{A}$ is diagonal. From the necessary condition for the optimality, we can derive a_ℓ as the eigenvector corresponding to the ℓ-th largest eigenvalue of the scatter matrix S.

Regression model. The PCA model can also be defined by using a regression model approach. In [123], the model was formulated based on the eigenvalue problem of correlation coefficient matrices by using normalized data matrix. Here, we derive the formulation that is equivalent to the Pearson's basic formulation. Let denote the j-th regression model as

$$x_{\cdot}^j = a^j f_{\cdot} + v^j, \tag{8.36}$$

where the j-th original variable $x_{\cdot}^j$ is the dependent variable, and latent variable $f_{\cdot}$ and its coefficient a^j are the independent variable and its corresponding regression coefficient, respectively. The goal is to minimize the sum of regression errors:

$$J_{\text{pca4}}^j = \sum_{k=1}^{N} \left(x_k^j - a^j f_k - v^j \right)^2. \tag{8.37}$$

Here, we can see that (8.37) is equivalent to a part of least squares criterion (8.32). Therefore, the constant term v^j is given as the mean value of $x_{\cdot}^j$.

Next, we search for the optimal independent variable $f_{\cdot}$ in a different way. The goal of regression analysis can also be regarded as maximization of the covariance between dependent variable and its regression value. In this context, we try to maximize the sum of squared covariance between the dependent variable $x_{\cdot}^j$ and the independent variable $f_{\cdot}$.

$$\begin{aligned} J_{\text{pca4}}(F) &= \sum_{j=1}^{p} \left(\sum_{k=1}^{N} (x_k^j - v^j)(f_k - \bar{f}) \right)^2 \\ &= \sum_{k=1}^{N} \sum_{\ell=1}^{N} \left((f_k - \bar{f})(f_\ell - \bar{f}) \sum_{j=1}^{p} (x_k^j - v^j)(x_\ell^j - v^j) \right), \end{aligned} \tag{8.38}$$

where $F = (f_1, \ldots, f_N)$. Under the condition of $\sum_{k=1}^{N} f_k^{\prime 2} = 1$, the optimal latent variable is derived by solving the following eigenvalue problem.

$$\hat{\mathsf{X}}\hat{\mathsf{X}}^\top \tilde{f} = \gamma \tilde{f}, \tag{8.39}$$

where $\hat{\mathsf{X}} = \mathsf{X} - 1_N v^\top$. The optimal $\tilde{f}$ is the N dimensional eigenvector corresponding to the maximum eigenvalue.

By the way, we can rewrite (8.39) as

$$\hat{\mathsf{X}}^\top \hat{\mathsf{X}}\hat{\mathsf{X}}^\top \tilde{f} = \gamma \hat{\mathsf{X}}^\top \tilde{f}. \tag{8.40}$$

By setting as $w = \hat{\mathsf{X}}^\top \tilde{f}$, (8.40) is given as

$$\hat{\mathsf{X}}^\top \hat{\mathsf{X}} w = \gamma w, \tag{8.41}$$

and is equivalent to the problem of finding the maximum eigenvalues of the scatter matrix S and their corresponding eigenvectors. In this sense, we can solve the PCA problem based on the regression model approach.

8.2.2 Local PCA Based on Fitting Low-Dimensional Subspace

In the remaining part of this section, samples are partitioned into c clusters (sub-models) and the degree of belongingness of sample k to model i is represented by membership u_{ki}. For deriving fuzzy partition, memberships are estimated under the following conditions:

$$u_{ki} \in [0, 1], \qquad i = 1, \ldots, c; k = 1, \ldots, N, \tag{8.42}$$

$$\sum_{i=1}^{c} u_{ki} = 1, \qquad k = 1, \ldots, N. \tag{8.43}$$

First, we try to estimate c different sub-models by partitioning samples into clusters. For the task of local PCA, the measure of fit is also used for clustering criterion, and the objective function is defined as the membership-weighted version of within-group-sum-of-errors:

$$J_{\mathrm{lpca1}}(U, V, A) = \sum_{i=1}^{c} \sum_{k=1}^{N} (u_{ki})^m \left(||x_k - v_i||^2 - \sum_{\ell=1}^{q} |a_{i\ell}^\top (x_k - v_i)|^2 \right), \tag{8.44}$$

where U, V and A are the total sets of memberships u_{ki}, cluster centers v_i and principal component vectors $a_{i\ell}$, respectively. $(u_{ki})^m$ is the weight for sample k in cluster i. The weighting exponent m is called "fuzzifier" and is used in the standard FCM clustering [6]. It plays a role for tuning the degree of fuzziness. The larger the m, the fuzzier the membership assignments. $a_{i\ell}, \ell = 1, \ldots, q$ are the basis vectors spanning the i-th subspace and v_i is the center of cluster i.

Here, (8.44) is equivalent to the objective function of the FCV clustering [6, 7]. The FCV clustering is the linear fuzzy clustering algorithm that uses q-dimensional linear varieties as prototypes of clusters. Therefore, it can be said that the FCV clustering is an extension of the PCA model based on fitting low-dimensional subspace.

The clustering algorithm is based on the alternate optimization technique. From the necessary condition $\partial J_{\mathrm{lpca1}}/\partial v_i = 0_p$, one of natural choice of the optimal cluster center is given as the membership weighted version of mean vector,

$$v_i = \frac{\sum_{k=1}^{N} (u_{ki})^m x_k}{\sum_{k=1}^{N} (u_{ki})^m}. \tag{8.45}$$

The optimal $a_{i\ell}$ are derived from necessary condition for the optimality $\partial J_{\mathrm{lpca1}}/\partial a_{i\ell} = 0_p$ and is the solution of the eigenvalue problem of generalized fuzzy scatter matrix,

$$S_i = \sum_{k=1}^{N} (u_{ki})^m (x_k - v_i)(x_k - v_i)^\top. \tag{8.46}$$

Because the optimal basis vectors $a_{i\ell}$ are the eigenvectors corresponding to the largest eigenvalues, the vectors can be regarded as fuzzy principal component vectors extracted in each cluster considering membership degrees [172].

Consequently, from condition $\partial J_{\mathrm{lpca1}}/\partial u_{ki} = 0$, memberships are updated as

$$u_{ki} = \left[\sum_{j=1}^{c} \left(\frac{D(x_k, P_i)}{D(x_k, P_j)} \right)^{\frac{1}{m-1}} \right]^{-1}, \tag{8.47}$$

where $D(x_k, P_i)$ is the squared distance between data point k and prototypical linear variety P_i,

$$D(x_k, P_i) = ||x_k - v_i||^2 - \sum_{\ell=1}^{q} |a_{i\ell}^\top (x_k - v_i)|^2. \tag{8.48}$$

When the goal is reduction of dimensionality, we estimate q dimensional projection $f_{ik} = (f_{ik}^1, \dots, f_{ik}^q)^\top$ in cluster i,

$$f_{ik}^\ell = a_{i\ell}^\top (x_k - v_i). \tag{8.49}$$

Using this latent variable, the optimal reconstruction of data vector x_k is given as $y_{ik} = v_i + \sum_{\ell=1}^{q} f_{ik}^\ell a_{i\ell}$.

8.2.3 Linear Clustering with Variance Measure of Latent Variables

Second, we use the variance of q dimensional projection $f_{ik} = (f_{ik}^1, \dots, f_{ik}^q)^\top$,

$$f_{ik}^\ell = a_{i\ell}^\top (x_k - v_i), \tag{8.50}$$

as the measure to be optimized in cluster i. Then, the objective function is given as follows:

$$J_{\text{lpca2}}(U, V, A) = \sum_{i=1}^{c} \sum_{k=1}^{N} (u_{ki})^m \left(\sum_{\ell=1}^{q} |a_{i\ell}^{\top}(x_k - v_i)|^2 \right). \tag{8.51}$$

Equation (8.51) is equivalent to the second term of (8.44). The objective function, however, cannot be used for the FCV clustering because the second derivative of (8.51) with respect to memberships u_{ki} is positive while the second term of (8.44) should be maximized. The FCM-type clustering algorithm can provide only the solution for minimization problems with respect to memberships u_{ki}.

If we consider minimization of (8.51), the solution is closely related to minor component analysis (MCA). MCA contrasts with PCA and tries to extract eigenvectors corresponding to the least eigenvalues. The first minor component is the normalized linear combination with minimum variance and the minor component vector is useful in estimating the orthonormal basis of subspace or noise subspace. For simple extraction of minor components, several algorithms that are mainly associated to neural networks have been proposed [93, 121].

Assume that $\alpha_{i\ell}$ is the ℓ-th eigenvector corresponding to the ℓ-th largest eigenvalue of fuzzy scatter matrix. The clustering criterion in the FCV clustering $D(x_k, P_i)$ is given as

$$\begin{aligned} D(x_k, P_i) &= ||x_k - v_i||^2 - \sum_{\ell=1}^{q} |\alpha_{i\ell}^{\top}(x_k - v_i)|^2 \\ &= \sum_{\ell=q+1}^{p} |\alpha_{i\ell}^{\top}(x_k - v_i)|^2. \end{aligned} \tag{8.52}$$

So, if $a_{i\ell}, \ell = 1, \ldots, q$ are given as $\alpha_{ip}, \alpha_{i,p-1}, \ldots, \alpha_{i,p-q+1}$, it can be said that minimization of (8.51) derives the same clustering result with that of the **FCV** algorithm using $p - q$ dimensional prototypes.

8.2.4 Local PCA Based on Lower Rank Approximation of Data Matrix

Third, another formulation of the FCV clustering is given by modifying the least squares criterion for principal component analysis. Introducing memberships u_{ki}, the least squares criterion for local PCA is defined as follows:

$$J_{\text{lpca3}}(U, V, F, A) = \sum_{i=1}^{c} \text{tr}\left\{ (\mathsf{X} - \mathsf{Y}_i)^{\top} \mathsf{U}_i^m (\mathsf{X} - \mathsf{Y}_i) \right\}, \tag{8.53}$$

where U_i is the $N \times N$ diagonal matrix $\mathsf{U}_i = \text{diag}(u_{i1}, \ldots, u_{iN})$. $\mathsf{Y}_i = (y_{i1}, \ldots, y_{iN})^{\top}$ is the lower rank approximation of data matrix X, which is estimated in cluster i as follows:

$$\mathsf{Y}_i = \mathsf{F}_i \mathsf{A}_i^{\top} + 1_N v_i^{\top}, \tag{8.54}$$

where $\mathsf{F}_i = (f_{i1}, \dots, f_{iN})^\top$ is the $N \times q$ score matrix and $\mathsf{A}_i = (a_{i1}, \dots, a_{iq})$ is the $p \times q$ principal component matrix. F and A are the sets of component score matrices and component matrices, respectively. 1_N is N dimensional vector whose all elements are 1.

With fixed memberships, the extraction of local principal components in each cluster is equivalent to the calculation of F_i, A_i and v_i such that the least squares criterion (8.53) is minimized.

From the necessary condition for the optimality of the objective function $\partial J_{\mathrm{lpca3}}/\partial v_i = 0_p$, we have

$$v_i = (1_N^\top \mathsf{U}_i^m 1_N)^{-1}(\mathsf{X}^\top - \mathsf{A}_i\mathsf{F}_i^\top)\mathsf{U}_i^m 1_N. \tag{8.55}$$

Usually, in PCA, the principal component score is constrained to have zero mean. Considering fuzzy partitioning, the constraint is replaced with $\mathsf{F}_i^\top \mathsf{U}_i^m 1_N = 0_q$. Then, (8.55) is reduced to

$$v_i = (1_N^\top \mathsf{U}_i^m 1_N)^{-1}\mathsf{X}^\top \mathsf{U}_i^m 1_N. \tag{8.56}$$

Equation (8.56) is equivalent to the updating rule for the cluster center v_i in the **FCV** algorithm (**FCM** algorithm). Substituting (8.54) and (8.56), the objective function (8.53) becomes

$$\begin{aligned} J_{\mathrm{lpca3}} = \sum_{i=1}^{c} \Big\{ & \mathrm{tr}(\mathsf{X}_i^\top \mathsf{U}_i^m \mathsf{X}_i) - 2\mathrm{tr}(\mathsf{X}_i^\top \mathsf{U}_i^m \mathsf{F}_i \mathsf{A}_i^\top) \\ & + \mathrm{tr}(\mathsf{A}_i \mathsf{F}_i^\top \mathsf{U}_i^m \mathsf{F}_i \mathsf{A}_i^\top) \Big\}, \end{aligned} \tag{8.57}$$

where $\mathsf{X}_i = \mathsf{X} - 1_N v_i^\top$. From $\partial J_{\mathrm{lpca3}}/\partial \mathsf{F}_i = O$,

$$\mathsf{F}_i \mathsf{A}_i^\top \mathsf{A}_i = \mathsf{X}_i \mathsf{A}_i. \tag{8.58}$$

Under the condition $\mathsf{A}_i^\top \mathsf{A}_i = I_q$, which is used in the **FCV** algorithm, we have $\mathsf{F}_i = \mathsf{X}_i \mathsf{A}_i$ and the objective function is transformed as follows:

$$\begin{aligned} J_{\mathrm{lpca3}} &= \sum_{i=1}^{c} \Big\{ \mathrm{tr}(\mathsf{X}_i^\top \mathsf{U}_i^m \mathsf{X}_i) - \mathrm{tr}(\mathsf{A}_i^\top \mathsf{X}_i^\top \mathsf{U}_i^m \mathsf{X}_i \mathsf{A}_i) \Big\} \\ &= J_{\mathrm{lpca1}}. \end{aligned} \tag{8.59}$$

Therefore it can be said that (8.53) is equivalent to the objective function of **FCV** and the minimization problem is solved by computing the q largest singular values of the generalized fuzzy scatter matrix and the associated vectors.

By the way, the objective function (8.53) can also be expressed as

$$J_{\mathrm{lpca3}}(U, V, F, A) = \sum_{i=1}^{c} \sum_{k=1}^{N} (u_{ki})^m \sum_{j=1}^{p} \left(x_k^j - \sum_{\ell=1}^{q} f_{ik}^{\ell} a_{i\ell}^j - v_i^j \right)^2. \tag{8.60}$$

This formulation means that the clustering criterion is composed of the component-wise approximation of data matrices.

8.2.5 Local PCA Based on Regression Model

Fourth, we expand the problem to finding c different regression models. Let denote the i-th regression model for variable j as

$$x^j_\cdot = a^j_i f_{i\cdot} + v^j_i, \tag{8.61}$$

where the j-th original variable $x^j_\cdot$ is the dependent variable, and latent variable $f_{i\cdot}$ and its coefficient a^j_i are the independent variable and its corresponding regression coefficient, respectively. The objective function for the i-th regression model is the membership-weighted sum of errors:

$$J^{ij}_{\mathrm{lpca4}} = \sum_{k=1}^{N} (u_{ki})^m \left(x^j_k - a^j_i f_{ik} - v^j_i \right)^2 . \tag{8.62}$$

Considering $\partial J^{ij}_{\mathrm{lpca4}} / \partial v^j_i = 0$, parameter v^j_i is given as

$$v^j_i = \frac{\sum_{k=1}^{N} (u_{ki})^m x^j_k}{\sum_{k=1}^{N} (u_{ki})^m} . \tag{8.63}$$

Here, the optimal latent variable should have the maximum covariance with the j-th original variable. So, we maximize the following squared sum of covariances ($\times \sum_{k=1}^{N} (u_{ki})^m$):

$$\begin{aligned} J^{i}_{\mathrm{lpca4}} &= \sum_{j=1}^{p} \left(\sum_{k=1}^{N} (u_{ki})^m (x^j_k - v^j_i)(f_{ik} - \bar{f}_i) \right)^2 \\ &= \sum_{k=1}^{N} \sum_{\ell=1}^{N} \Bigg((u_{ki})^m (u_{\ell i})^m (f_{ik} - \bar{f}_i)(f_{i\ell} - \bar{f}_i) \\ &\quad \times \sum_{j=1}^{p} (x^j_k - v^j_i)(x^j_\ell - v^j_i) \Bigg) . \end{aligned} \tag{8.64}$$

Under the condition that $\sum_{k=1}^{N} (u_{ki})^m (f_{ik})^2 = 1$, the Lagrangian function is derived as

$$L^{i}_{\mathrm{lpca4}} = J^{i}_{\mathrm{lpca4}} - \gamma_i \left(\sum_{k=1}^{N} (u_{ki})^m (f_{ik})^2 - 1 \right), \tag{8.65}$$

where γ_i is the Lagrange multiplier. From $\partial L^{i}_{\mathrm{lpca4}} / \partial f_{ik} = 0$, the necessary condition is

$$\sum_{j=1}^{p} \sum_{\ell=1}^{N} (u_{ki})^m (u_{\ell i})^m (x^j_k - v^j_i)(x^j_\ell - v^j_i) f_{i\ell} - \gamma_i (u_{ki})^m f_{ik} = 0, \tag{8.66}$$

and $\sum_{\ell=1}^{N}(u_{\ell i})^m(x_\ell^j - v_i^j) = 0$ gives $\gamma_i = 0$ and

$$f_{ik} = \frac{1}{\sqrt{\sum_{\ell=1}^{N}(u_{\ell i})^m}}, \quad k = 1, \ldots, N. \tag{8.67}$$

By the way, (8.66) is equivalent to

$$\mathsf{U}_i^{m/2}\mathsf{X}_i\mathsf{X}_i^\top\mathsf{U}_i^{m/2}(\mathsf{U}_i^{m/2}\tilde{f}_i) = \gamma_i(\mathsf{U}_i^{m/2}\tilde{f}_i), \tag{8.68}$$

where $\mathsf{X}_i = \mathsf{X} - 1_N v_i^\top$, $\mathsf{U}_i = \mathrm{diag}(u_{1i}, \ldots, u_{Ni})$ and $\tilde{f}_i = (f_{i1}, \ldots, f_{iN})^\top$. So, γ_i and $\mathsf{U}_i^{m/2}\tilde{f}_i$ are the solution of the eigenvalue problem of $\mathsf{U}_i^{m/2}\mathsf{X}_i\mathsf{X}_i^\top\mathsf{U}_i^{m/2}$.

Here, it must be noted that the eigenvalue of $\mathsf{U}_i^{m/2}\mathsf{X}_i\mathsf{X}_i^\top\mathsf{U}_i^{m/2}$ is equivalent to that of $\mathsf{X}_i^\top\mathsf{U}_i^m\mathsf{X}_i$. Multiplying both members by $\mathsf{X}_i^\top\mathsf{U}_i^{m/2}$, the characteristic equation (8.68) is arranged as

$$\mathsf{X}_i^\top\mathsf{U}_i^m\mathsf{X}_i\mathsf{X}_i^\top\mathsf{U}_i^{m/2}(\mathsf{U}_i^{m/2}\tilde{f}_i) = \gamma_i(\mathsf{X}_i^\top\mathsf{U}_i^m\tilde{f}_i). \tag{8.69}$$

So, the equation can be rewritten as

$$\mathsf{X}_i^\top\mathsf{U}_i^m\mathsf{X}_i w_i = \gamma_i w_i, \tag{8.70}$$

where $w_i = \mathsf{X}_i^\top\mathsf{U}_i^m\tilde{f}_i$, and w_i is the eigenvector of generalized fuzzy scatter matrix S_i. Therefore, latent variable f_i. is closely related to that of the FCV clustering.

On the other hand, membership u_{ki} cannot be derived from the minimization of (8.64) because (8.64) is to be maximized. Then, we use the following objective function:

$$J_{\mathrm{lpca4}}(U, V, F) = \sum_{i=1}^{c}\sum_{j=1}^{p}\left\{\sum_{k=1}^{N}(u_{ki})^m(x_k^j - v_i^j)^2 - \left(\sum_{k=1}^{N}(u_{ki})^m(x_k^j - v_i^j)(f_{ik} - \bar{f}_i)\right)^2\right\} \tag{8.71}$$

Because the second term of the objective function (8.71) includes the product of $(u_{ki})^m$ and $(u_{\ell i})^m$ $(k \neq \ell)$, new memberships should be calculated by a technique used in relational clustering [50, 137], in which one of two memberships is fixed. With fixed $(u_{\ell i})^m$, the clustering criterion $D(x_k, P_i)$ is given as

$$\begin{aligned} D(x_k, P_i) &= \|x_k - v_i\|^2 - \sum_{\ell=1}^{N}(u_{\ell i})^m f_{ik} f_{i\ell}(x_k - v_i)^\top(x_\ell - v_i) \\ &= \|x_k - v_i\|^2 - f_{ik}(x_k - v_i)^\top\left(\sum_{\ell=1}^{N}(u_{\ell i})^m f_{i\ell}(x_\ell - v_i)\right) \\ &= \|x_k - v_i\|^2 - f_{ik}(x_k - v_i)^\top\mathsf{X}_i^\top\mathsf{U}_i^m\tilde{f}_i \\ &= \|x_k - v_i\|^2 - f_{ik}(x_k - v_i)^\top w_i \\ &= \|x_k - v_i\|^2 - |a_i(x_k - v_i)|^2. \end{aligned} \tag{8.72}$$

Therefore, it can be said that the same clustering result with the FCV clustering is derived from the regression model of (8.61). In this way, linear fuzzy clustering is a technique for estimating meaningful latent variables.

8.3 Fuzzy Clustering-Based Local Quantification of Categorical Variables

Homogeneity analysis [13, 41] is a quantification technique for representing the structure of non-numerical multivariate data and tries to minimize departures from perfect homogeneity measured by the Gifi loss function. Minimization of the Gifi loss function is based on approximation of a quantified data matrix, so the algorithm is similar to that of PCA with least squares criterion.

In this section, we introduce a local quantification method that can be regarded as a linear fuzzy clustering technique for categorical variables.

8.3.1 Homogeneity Analysis

Suppose that we have collected data on N objects on p categorical variables with $K_j, j = 1, \dots, p$ categories. The categories of each variable are often nominal, i.e., only the classes formed by the objects play a role. These non-numerical variables are represented by indicator matrices. Let G_j denote the $N \times K_j$ indicator matrix corresponding to variable j and its entries be the binary variables as follows:

$$g_{kjl} = \begin{cases} 1 & ; \quad \text{if object } k \text{ belongs to category } l. \\ 0 & ; \quad \text{otherwise.} \end{cases}$$

$$\mathsf{G}_j = \begin{pmatrix} g_{1j1} & \cdots & g_{1jl} & \cdots & g_{1jK_j} \\ \vdots & \ddots & \vdots & \ddots & \vdots \\ g_{kj1} & \cdots & g_{kjl} & \cdots & g_{kjK_j} \\ \vdots & \ddots & \vdots & \ddots & \vdots \\ g_{Nj1} & \cdots & g_{Njl} & \cdots & g_{NjK_j} \end{pmatrix}.$$

These matrices can be collected in an $N \times K$ partitioned matrix $\mathsf{G} = [\mathsf{G}_1, \mathsf{G}_2, \dots, \mathsf{G}_p]$, where $K = \sum_{j=1}^{p} K_j$ is the total number of categories.

The goal of quantification of categorical data is to represent these objects in a q dimensional space ($q < p$). Homogeneity analysis [13, 41] is a basic technique of non-linear multivariate analysis and aims at the representation of the structure of non-numerical multivariate data by assigning scores to objects and the categories of variables. Let W_j denote the $K_j \times q$ matrix containing the multiple category quantification of variable j and Z be an $N \times q$ matrix containing the resulting q object scores as follows:

$$\mathsf{W}_j = \begin{pmatrix} w_{j1}^1 & w_{j1}^2 & \cdots & w_{j1}^q \\ w_{j2}^1 & w_{j2}^2 & \cdots & w_{j2}^q \\ \vdots & \vdots & \ddots & \vdots \\ w_{jK_j}^1 & w_{jK_j}^2 & \cdots & w_{jK_j}^q \end{pmatrix}, \tag{8.73}$$

$$\mathsf{Z} = \begin{pmatrix} z_1^1 & z_1^2 & \cdots & z_1^q \\ z_2^1 & z_2^2 & \cdots & z_2^q \\ \vdots & \vdots & \ddots & \vdots \\ z_N^1 & z_N^2 & \cdots & z_N^q \end{pmatrix}. \tag{8.74}$$

Homogeneity analysis is based on the principle that a scale consisting of nominal variables is homogenizable if all variables can be quantified in such a way that the resulting scale is homogeneous, i.e., all the variables in the scale are linearly related. The departures from perfect homogeneity are measured by the Gifi loss function.

$$J_{\text{ha}}(\mathsf{Z}, W) = \frac{1}{p}\sum_{j=1}^{p} \text{tr}\{(\mathsf{Z} - \mathsf{G}_j\mathsf{W}_j)^\top(\mathsf{Z} - \mathsf{G}_j\mathsf{W}_j)\}, \tag{8.75}$$

where $W = (\mathsf{W}_1, \ldots, \mathsf{W}_p)$. In order to avoid the trivial solution, the loss function is minimized under the conditions,

$$1_N^\top \mathsf{Z} = 0_q^\top, \tag{8.76}$$

$$\mathsf{Z}^\top\mathsf{Z} = NI_q, \tag{8.77}$$

where 1_N is the N dimensional vector whose all elements are 1, 0_q is the q dimensional zero-vector and I_q is the $q \times q$ unit matrix.

Here, the objective of choosing scores and weights is one of the possible definitions of finding the principal components. If the rank of the score matrix W_j is constrained to be 1, the matrix can be decomposed into

$$\mathsf{W}_j = q_j\tilde{a}_j^\top, \tag{8.78}$$

where q dimensional vector $\tilde{a}_j$ consists of the j-th elements of the principal component vectors and q_j is the weight vector arranging the categories on the principal subspace. Substituting (8.78), the Gifi loss function is transformed into

$$J_{\text{ha}} = \frac{1}{p}\sum_{j=1}^{p} \text{tr}\{(\mathsf{Z} - \mathsf{G}_j q_j\tilde{a}_j^\top)^\top(\mathsf{Z} - \mathsf{G}_j q_j\tilde{a}_j^\top)\}. \tag{8.79}$$

When the weight vector q_j satisfies the conditions that $1_N^\top \mathsf{G}_j q_j = 0$ and $q_j^\top \mathsf{G}_j^\top \mathsf{G}_j q_j = N$, the loss function can be replaced with

$$J_{\text{ha}} = \frac{1}{p}\sum_{j=1}^{p} \text{tr}\{(Z\tilde{a}_j - \mathsf{G}_j q_j)^\top (Z\tilde{a}_j - \mathsf{G}_j q_j)\}. \tag{8.80}$$

Because $\mathsf{G}_j q_j$ is the numerical representation of categorical data matrix, the loss function (8.80) is closely related to the least squares criterion for PCA based on the component-wise approximation. In this sense, homogeneity analysis can be regarded as a technique of PCA for categorical data [41].

8.3.2 Local Quantification Method and FCV Clustering of Categorical Data

In this subsection, we consider simultaneous application of fuzzy clustering and homogeneity analysis that can be regarded as local homogeneity analysis. The goal is to partition objects into c clusters estimating category quantification W_{ij} and object score matrix Z_i in each cluster [60]. Fuzzy partitioning of N objects is performed by introducing memberships $u_{ki}, i = 1, \ldots, c; k = 1, \ldots, N$ and u_{ki} denotes the membership degree of object k to cluster i. The objective function is defined as follows:

$$J_{\text{lha1}}(U, Z, W) = \frac{1}{p}\sum_{i=1}^{c}\sum_{j=1}^{p} \text{tr}\{(\mathsf{Z}_i - \mathsf{G}_j\mathsf{W}_{ij})^\top \mathsf{U}_i^m (\mathsf{Z}_i - \mathsf{G}_j\mathsf{W}_{ij})\}, \tag{8.81}$$

where $Z = (\mathsf{Z}_1, \ldots, \mathsf{Z}_c)$ and $W = (\mathsf{W}_{11}, \ldots, \mathsf{W}_{cp})$. To derive unique solution, the objective function (8.81) is minimized under the following conditions:

$$u_i^\top \mathsf{Z}_i = 0_q^\top, \tag{8.82}$$

$$\mathsf{Z}_i^\top \mathsf{U}_i \mathsf{Z}_i = \Big\{\sum_{k=1}^{N} u_{ki}\Big\} I_q, \tag{8.83}$$

where $u_i = (u_{1i}, u_{2i}, \ldots, u_{Ni})^\top$.

The algorithm can be written as follows:

Algorithm LHAO: *Local Homogeneity Analysis Considering Partition of Objects* [60]

LHAO1. [Generate initial value:] Generate initial values for $\bar{Z} = (\bar{\mathsf{Z}}_1, \ldots, \bar{\mathsf{Z}}_c)$ and $\bar{W} = (\bar{\mathsf{W}}_{11}, \ldots, \bar{\mathsf{W}}_{cp})$, and normalize them so that they satisfy the constraints (8.82) and (8.83).

LHAO2. [Find optimal U:] Calculate

$$\bar{U} = \arg\min_{U \in \mathcal{U}_f} J_{\text{lha1}}(U, \bar{Z}, \bar{W}). \tag{8.84}$$

LHAO3. [Find optimal Z:] Calculate

$$\bar{Z} = \arg\min_{Z} J_{\mathrm{lha1}}(\bar{U}, Z, \bar{W}), \tag{8.85}$$

and normalize them so that they satisfy the constraints (8.82) and (8.83).

LHAO4. [Find optimal W:] Calculate

$$\bar{W} = \arg\min_{W} J_{\mathrm{lha1}}(\bar{U}, \bar{Z}, W). \tag{8.86}$$

LHAO5. [Test convergence:] If all parameters are convergent, stop; else go to LHAO2.

End LHAO.

The optimal solution is derived based on the iterative least squares technique. Let O be a zero-matrix. From the necessary condition for the optimality $\partial J_{\mathrm{lha1}}/\partial \mathsf{W}_{ij} = O$, the updating rule for W_{ij} is derived as

$$\mathsf{W}_{ij} = D_{ij}^{-1}\mathsf{G}_j^\top \mathsf{U}_i^m \mathsf{Z}_i, \tag{8.87}$$

where $D_{ij} = \mathsf{G}_j^\top \mathsf{U}_i^m \mathsf{G}_j$. Consequently, from $\partial J_{\mathrm{lha1}}/\partial \mathsf{Z}_i = O$ and $\partial J_{\mathrm{lha1}}/\partial u_{ki} = 0$, we have

$$\mathsf{Z}_i = \sum_{j=1}^{p} \mathsf{G}_j \mathsf{W}_{ij}, \tag{8.88}$$

and

$$u_{ki} = \left[\sum_{j=1}^{c} \left(\frac{D(g_k, \mathsf{Z}_i; \mathsf{W}_i)}{D(g_k, \mathsf{Z}_j; \mathsf{W}_j)} \right)^{\frac{1}{m-1}} \right]^{-1}, \tag{8.89}$$

where $g_k = (g_{k1}^\top, \ldots, g_{kp}^\top)^\top$ is the K dimensional response vector on object k and $\mathsf{W}_i = [\mathsf{W}_{i1}, \ldots, \mathsf{W}_{ip}]$. $D(g_k, \mathsf{Z}_i; \mathsf{W}_i)$ is the departure of object k from perfect homogeneity in cluster i,

$$D(g_k, \mathsf{Z}_i; \mathsf{W}_i) = \sum_{j=1}^{p} \sum_{t=1}^{q} \left(z_{ik}^t - \sum_{\ell=1}^{K_j} g_{kj\ell} w_{ij\ell}^t \right)^2. \tag{8.90}$$

Next, we show the close relation between local homogeneity analysis and the FCV clustering. If the rank of the score matrix W_{ij} is constrained to be 1, the matrix can be decomposed into

$$\mathsf{W}_{ij} = q_{ij} \tilde{a}_{ij}^\top, \tag{8.91}$$

where q dimensional vector $\tilde{a}_{ij}$ consists of the j-th elements of the basis vectors and q_{ij} is the weight vector arranging the categories on the prototypical linear variety of cluster i. Substituting (8.91), the objective function is transformed into

$$J_{\mathrm{lha1}} = \frac{1}{p} \sum_{i=1}^{c} \sum_{j=1}^{p} \mathrm{tr}\{(\mathsf{Z}_i - \mathsf{G}_j q_{ij} \tilde{a}_{ij}^\top)^\top \mathsf{U}_i^m (\mathsf{Z}_i - \mathsf{G}_j q_{ij} \tilde{a}_{ij}^\top)\}. \tag{8.92}$$

When the weight vector q_{ij} satisfies the conditions that $1_N^\top \mathsf{U}_i^m \mathsf{G}_j q_{ij} = 0$ and $q_{ij}^\top \mathsf{G}_j^\top \mathsf{U}_i^m \mathsf{G}_j q_j = 1_N^\top \mathsf{U}_i^m 1_N$, the objective function can be replaced with

$$J_{\text{lha1}} = \frac{1}{p} \sum_{i=1}^{c} \sum_{j=1}^{p} \text{tr}\{(\mathsf{Z}_i \tilde{a}_{ij} - \mathsf{G}_j q_{ij})^\top \mathsf{U}_i^m (\mathsf{Z}_i \tilde{a}_{ij} - \mathsf{G}_j q_{ij})\}. \quad (8.93)$$

Because $\mathsf{G}_j q_{ij}$ is the numerical representation of categorical data matrix in cluster i, (8.93) is closely related to the least squares criterion for **FCV** based on the component-wise approximation. In this sense, this local homogeneity analysis formulation can be regarded as a technique of **FCV** for categorical data.

8.3.3 Application to Classification of Variables

Another possible local homogeneity analysis is classification of variables based on FCM-type clustering [157]. Fuzzy partitioning of p variables is performed by introducing memberships $u_{ji}, i = 1, \ldots, c; j = 1, \ldots, p$ and u_{ji} denotes the membership degree of variable j to cluster i. The objective function is defined as follows:

$$J_{\text{lha2}}(U, Z, W) = \frac{1}{p} \sum_{i=1}^{c} \sum_{j=1}^{p} u_{ji} \text{tr}\{(\mathsf{Z}_i - \mathsf{G}_j \mathsf{W}_{ij})^\top (\mathsf{Z}_i - \mathsf{G}_j \mathsf{W}_{ij})\}. \quad (8.94)$$

To derive unique solution, the objective function (8.94) is minimized under the following conditions:

$$1_N^\top \mathsf{Z}_i = 0_q^\top, \quad (8.95)$$

$$\mathsf{Z}_i^\top \mathsf{Z}_i = N I_q. \quad (8.96)$$

From the necessary condition for the optimality $\partial J_{\text{lha2}} / \partial \mathsf{W}_{ij} = O$, the updating rule for W_{ij} is derived as

$$\mathsf{W}_{ij} = D_{ij}^{-1} \mathsf{G}_j^\top \mathsf{Z}_i, \quad (8.97)$$

where $D_{ij} = \mathsf{G}_j^\top \mathsf{G}_j$. Consequently, from $\partial J_{\text{lha2}} / \partial \mathsf{Z}_i = O$ and $\partial J_{\text{lha2}} / \partial u_{ji} = 0$, we have

$$\mathsf{Z}_i = \sum_{j=1}^{p} (u_{ji})^m \mathsf{G}_j \mathsf{W}_{ij}, \quad (8.98)$$

and

$$u_{ji} = \left[\sum_{\ell=1}^{c} \left(\frac{D(G_j, \mathsf{Z}_i; \mathsf{W}_i)}{D(G_j, \mathsf{Z}_\ell; \mathsf{W}_\ell)} \right)^{\frac{1}{m-1}} \right]^{-1}. \quad (8.99)$$

$D(G_j, \mathsf{Z}_i; \mathsf{W}_i)$ is the departure of variable j from perfect homogeneity in cluster i,

$$D(G_j, \mathsf{Z}_i; \mathsf{W}_i) = \text{tr}\{(\mathsf{Z}_i - \mathsf{G}_j \mathsf{W}_{ij})^\top (\mathsf{Z}_i - \mathsf{G}_j \mathsf{W}_{ij})\}. \quad (8.100)$$

8.3.4 An Illustrative Example

We show an illustrative example of local quantification problems. Table 8.1 shows the indicator matrix composed of responses by 10 objects (responder) for 15 variables (questions). This matrix can also be represented like Table 8.2 by arranging the order of variables. The object 1-5 are linearly related to the variables as is noted in bold in Table 8.1 while the object 6-10 are also linearly related to the variables like Table 8.2. But the two object groups have no relations. So, we can capture the linear relations only when we partition the objects into two clusters.

Table 8.3 shows the comparison of memberships derived by two different local quantification methods. "(a)u_{ki}" is the membership derived by local homogeneity analysis considering partition of objects [60] with $q = 1$, $c = 2$ and $m = 2$. u_{ki} represents the membership degree of object k in cluster i. We can see that the local homogeneity analysis method captured local linear relation by partitioning objects into two appropriate clusters. On the other hand, "(b)u_{ji}" is the membership derived by local homogeneity analysis for classification of variables [157] with same parameters, in which objects and variables were interchanged with

Table 8.1. Artificial data set to be analyzed by partitioning objects

	Variable														
Object	1	2	3	4	5	6	7	8	9	10	11	12	13	14	15
1	**1**	**2**	**2**	**2**	**2**	**2**	**3**	**3**	**3**	**3**	**3**	**1**	**1**	**1**	**1**
2	**1**	**1**	**2**	**2**	**2**	**2**	**2**	**3**	**3**	**3**	**3**	**3**	**1**	**1**	**1**
3	**1**	**1**	**1**	**2**	**2**	**2**	**2**	**2**	**3**	**3**	**3**	**3**	**3**	**1**	**1**
4	**1**	**1**	**1**	**1**	**2**	**2**	**2**	**2**	**2**	**3**	**3**	**3**	**3**	**3**	**1**
5	**1**	**1**	**1**	**1**	**1**	**2**	**2**	**2**	**2**	**2**	**3**	**3**	**3**	**3**	**3**
6	3	3	3	3	3	1	1	1	1	1	2	2	2	2	2
7	2	3	3	3	3	1	1	1	1	3	2	2	2	2	1
8	2	3	2	3	3	1	3	1	1	3	1	2	2	2	1
9	2	3	2	3	2	1	3	1	3	3	1	2	2	1	1
10	2	3	2	2	2	3	3	1	3	3	1	2	1	1	1

Table 8.2. Arranged matrix of Table 8.1

	Variable														
Object	8	15	11	14	13	12	1	3	5	4	2	10	7	9	6
1	3	1	3	1	1	1	1	2	2	2	2	3	3	3	2
2	3	1	3	1	1	3	1	2	2	2	1	3	2	3	2
3	2	1	3	1	3	3	1	1	2	2	1	3	2	3	2
4	2	1	3	3	3	3	1	1	2	1	1	3	2	2	2
5	2	3	3	3	3	3	1	1	1	1	1	2	2	2	2
6	**1**	**2**	**2**	**2**	**2**	**2**	**3**	**3**	**3**	**3**	**3**	**1**	**1**	**1**	**1**
7	**1**	**1**	**2**	**2**	**2**	**2**	**2**	**3**	**3**	**3**	**3**	**3**	**1**	**1**	**1**
8	**1**	**1**	**1**	**2**	**2**	**2**	**2**	**2**	**3**	**3**	**3**	**3**	**3**	**1**	**1**
9	**1**	**1**	**1**	**1**	**2**	**2**	**2**	**2**	**2**	**3**	**3**	**3**	**3**	**3**	**1**
10	**1**	**1**	**1**	**1**	**1**	**2**	**2**	**2**	**2**	**2**	**3**	**3**	**3**	**3**	**3**

Table 8.3. Comparison of memberships

Object (Variable)	(a) u_{ki}		(b) u_{ji}	
	i=1	i=2	i=1	i=2
1	1.00	0.00	0.34	0.66
2	1.00	0.00	0.87	0.13
3	1.00	0.00	0.96	0.04
4	1.00	0.00	0.77	0.23
5	1.00	0.00	0.06	0.94
6	0.00	1.00	0.06	0.94
7	0.00	1.00	0.32	0.68
8	0.00	1.00	0.34	0.66
9	0.00	1.00	0.20	0.80
10	0.00	1.00	0.14	0.86

variables and objects, respectively. Then, u_{ji} represents the membership degree of variable j in cluster i. Note that the interchanged model failed to capture the local substructures. It is because, in the interchanged data matrix, the mutual linear relations among objects in the two clusters do not have clear difference. In this way, the two local homogeneity analysis models are not interchangeable and we should use the appropriate model for each problem.

9 Extended Algorithms for Local Multivariate Analysis

In real application, we often face the difficulties such as handling missing data, noise or outliers, incorporating prior knowledge and integrating multiple algorithms. In this chapter, we introduce several extended algorithms for local multivariate analysis.

9.1 Clustering of Incomplete Data

Missing values have frequently been encountered in data analysis in real applications. There are many approaches to handle data sets including missing values. In this section, we briefly review several approaches to the FCM clustering of incomplete data and introduce techniques for the FCV clustering that can handle missing values.

9.1.1 FCM Clustering of Incomplete Data Including Missing Values

Unfortunately, there is no general method for dealing with missing values in fuzzy clustering. In this subsection, we review the four strategies introduced by Hathaway and Bezdek [49].

Whole data strategy (WDS). When we have only a few incomplete samples including missing values, we can adopt the whole data strategy (WDS), in which all incomplete samples are deleted and the **FCM** algorithm is applied to the remaining complete data. Although this strategy is very simple and easy to apply, it is obvious that we cannot derive reliable solutions for data with many incomplete samples because elimination brings a loss of information.

Partial distance strategy (PDS). If we want to calculate the distance between an incomplete sample and a cluster center, a partial distance is available. The partial distance strategy (PDS) was used in pattern recognition [26], and has been applied to the FCM clustering [100, 155].

Partial (squared Euclidean) distances are calculated using all available (observed) feature values, and are sometime weighted by the reciprocal of the

S. Miyamoto et al.: Algorithms for Fuzzy Clustering, STUDFUZZ 229, pp. 195–233, 2008.
springerlink.com

proportion of components used. Assume that we have a binary variable h_k^j that is defined by

$$h_k^j = \begin{cases} 1 & ; x_k^j \text{ is observed.} \\ 0 & ; x_k^j \text{ is missing.} \end{cases}$$

Then, the partial squared distance between sample k and the center of cluster i is given as

$$D(x_k, v_i) = \sum_{j=1}^{p} h_k^j (x_k^j - v_i^j)^2. \tag{9.1}$$

Optimal completion strategy (OCS). Another approach is to impute missing values. This strategy has close relation with the EM algorithm [25], in which missing values are imputed by maximum likelihood estimates in an iterative optimization procedure. In this approach, missing values are regarded as additional parameters to be optimized in the clustering algorithm, i.e., missing elements are replaced with their estimates so that the value of the objective function is minimized in each iteration step. From the necessary condition for the optimality, the estimate of a missing element x_k^j is calculated by considering the current fuzzy c-partition and cluster centers as

$$x_k^j = \frac{\sum_{i=1}^{c} (u_{ki})^m v_i^j}{\sum_{i=1}^{c} (u_{ki})^m}. \tag{9.2}$$

In the case of the entropy based method, we put $m = 1$. The estimate (9.2) is the membership-weighted average of v_i^j. Namely, the estimate might take a value that are far from all cluster centers when the sample has ambiguous memberships, and errors in imputation may accumulate and propagate through iterative optimization procedures.

Nearest prototype strategy (NPS). The fourth strategy is a simple modification of OCS, in which missing elements are imputed considering only the nearest prototype. Here, the "nearest prototype" means the prototype whose partial distance is minimum. When we note the partial distance between sample k and the center of cluster i as $D(x_k, v_i)$, the estimate of a missing element x_k^j is given as

$$x_k^j = v_\phi^j \quad \text{where } D(x_k, v_\phi) = \min\{D(x_k, v_1), \ldots, D(x_k, v_c)\}. \tag{9.3}$$

This strategy is somewhat similar with the classification EM algorithm (CEM algorithm) [16] that performs a classification version of the EM algorithm using a maximum a posteriori principle, which converts u_{ki} to a discrete classification (0 or 1).

9.1.2 Linear Fuzzy Clustering with Partial Distance Strategy

Several methods that extract principal components without elimination or imputation of data have been proposed [146, 148, 168]. Shibayama [146] proposed a PCA (principal component analysis)-like method to capture the structure of incomplete multivariate data without any imputations and statistical assumptions. The method is derived using lower rank approximation of a data matrix including missing values, which accomplishes minimization of the least squares criterion.

In this subsection, we enhance the method to partition an incomplete data set including missing values into several linear fuzzy clusters using least squares criterion. To handle missing values, we use the component-wise formulation of the FCV clustering criterion (8.53) and minimize only the deviations between x_k^j and y_{ik}^j where x_k^j are observed, and y_{ik}^j corresponding to missing values are determined incidentally.

The objective function to be minimized is the weighted-sum of partial squared distances,

$$J_{\text{fcv-pds}}(U, V, F, A) = \sum_{i=1}^{c}\sum_{k=1}^{N}(u_{ki})^m \sum_{j=1}^{p} h_k^j \left(x_k^j - \sum_{\ell=1}^{q} f_{ik}^{\ell} a_{i\ell}^j - v_i^j\right)^2. \tag{9.4}$$

For the entropy-based method, it is given as follows:

$$J_{\text{efcv-pds}}(U, V, F, A) = \sum_{i=1}^{c}\sum_{k=1}^{N} u_{ki} \sum_{j=1}^{p} h_k^j \left(x_k^j - \sum_{\ell=1}^{q} f_{ik}^{\ell} a_{i\ell}^j - v_i^j\right)^2 + \nu \sum_{i=1}^{c}\sum_{k=1}^{N} u_{ki}\log u_{ki}, \tag{9.5}$$

where F and A are the sets of component score matrices and component matrices, respectively. We can use either $J(U, V, F, A) = J_{\text{fcv-pds}}(U, V, F, A)$ or $J(U, V, F, A) = J_{\text{efcv-pds}}(U, V, F, A)$.

To obtain a unique solution, the objective function is minimized under the constrains that

$$\mathsf{F}_i^\top \mathsf{U}_i^m \mathsf{F}_i = I_q \quad ; i = 1, \ldots, c, \tag{9.6}$$

$$\mathsf{F}_i^\top \mathsf{U}_i^m 1_N = 0_q^\top \quad ; i = 1, \ldots, c, \tag{9.7}$$

$$\sum_{i=1}^{c} u_{ki} = 1 \quad ; k = 1, \ldots, N, \tag{9.8}$$

and $\mathsf{A}_i^\top \mathsf{A}_i$ is diagonal. I_q is $q \times q$ unit matrix and $0_q^\top$ is the q dimensional zero vector. 1_N is the N dimensional vector whose elements are all 1. In the case of the entropy-based method, we put $m = 1$. These constraints are somewhat different from those of the FCV clustering but they are common in statistic literature.

The algorithm for the FCV clustering with partial distance strategy can be written as follows [54]:

Algorithm FCV-PDS: *Fuzzy c-Varieties with Partial Distance Strategy*

FCV-PDS1. [Generate initial value:] Generate initial values for $\bar{U}=(\bar{\mathsf{U}}_1,\dots,\bar{\mathsf{U}}_c)$, $\bar{V}=(\bar{v}_1,\dots,\bar{v}_c)$, $\bar{F}=(\bar{\mathsf{F}}_1,\dots,\bar{\mathsf{F}}_c)$ and $\bar{A}=(\bar{\mathsf{A}}_1,\dots,\bar{\mathsf{A}}_c)$, and normalize them so that they satisfy the constraints (9.6)-(9.8) and $\bar{\mathsf{A}}_i^\top\bar{\mathsf{A}}_i$ is diagonal.
FCV-PDS2. [Find optimal U:] Calculate

$$\bar{U}=\arg\min_{U\in\mathcal{U}_f} J(U,\bar{V},\bar{F},\bar{A}). \tag{9.9}$$

FCV-PDS3. [Find optimal V:] Calculate

$$\bar{V}=\arg\min_{V} J(\bar{U},V,\bar{F},\bar{A}). \tag{9.10}$$

FCV-PDS4. [Find optimal F:] Calculate

$$\bar{F}=\arg\min_{F} J(\bar{U},\bar{V},F,\bar{A}), \tag{9.11}$$

and normalize them so that they satisfy the constraints (9.6) and (9.7).
FCV-PDS5. [Find optimal A:] Calculate

$$\bar{A}=\arg\min_{A} J(\bar{U},\bar{V},\bar{F},A), \tag{9.12}$$

and transform them so that each $\mathsf{A}_i^\top\mathsf{A}_i$ is diagonal.
FCV-PDS6. [Test convergence:] If all parameters are convergent, stop; else go to FCV-PDS2.
End FCV-PDS.

The orthogonal matrices in FCV-PDS1, FCV-PDS4 and FCV-PDS5 are obtained by such a technique as Gram-Schmidt's orthgonalization.

To derive the optimal A and V, we rewrite the objective function (9.4) as follows:

$$J_{\text{fcv-pds}}=\sum_{i=1}^{c}\sum_{j=1}^{p}(\tilde{x}_j-\mathsf{F}_i\tilde{a}_i^j-1_N v_i^j)^\top\mathsf{U}_i^m\tilde{\mathsf{H}}_j(\tilde{x}_j-\mathsf{F}_i\tilde{a}_i^j-1_N v_i^j), \tag{9.13}$$

where

$$\begin{aligned}\mathsf{X}&=(\tilde{x}_1,\dots,\tilde{x}_p),\\ \mathsf{A}_i&=(\tilde{a}_i^1,\dots,\tilde{a}_i^p)^\top,\\ \tilde{\mathsf{H}}_j&=\operatorname{diag}(h_1^j,\dots,h_N^j).\end{aligned}$$

Let 0_q be the q dimensional zero vector whose length is zero. From $\partial J_{\text{fcv-pds}}/\partial \tilde{a}_i^j = 0_q$ and $\partial J_{\text{fcv-pds}}/\partial v_i^j = 0$, we have

$$\tilde{a}_i^j = (\mathsf{F}_i^\top \mathsf{U}_i^m \tilde{\mathsf{H}}_j \mathsf{F}_i)^{-1} \mathsf{F}_i^\top \mathsf{U}_i^m \tilde{\mathsf{H}}_j (\tilde{x}_j - 1_N v_i^j), \tag{9.14}$$

$$v_i^j = (1_N^\top \mathsf{U}_i^m \tilde{\mathsf{H}}_j 1_N)^{-1} 1_N^\top \mathsf{U}_i^m \tilde{\mathsf{H}}_j (\tilde{x}_j - \mathsf{F}_i \tilde{a}_i^j). \tag{9.15}$$

In the same way, we can derive the optimal F. The objective function (9.4) is equivalent to

$$J_{\text{fcv-pds}} = \sum_{i=1}^{c} \sum_{k=1}^{N} (u_{ki})^m (x_k - \mathsf{A}_i f_{ik} - v_i)^\top \mathsf{H}_k (x_k - \mathsf{A}_i f_{ik} - v_i), \tag{9.16}$$

and $\partial J_{\text{fcv-pds}}/\partial f_{ik} = 0_q$ yields

$$f_{ik} = (\mathsf{A}_i^\top \mathsf{H}_k \mathsf{A}_i)^{-1} \mathsf{A}_i^\top \mathsf{H}_k (x_k - v_i), \tag{9.17}$$

where $\mathsf{H}_k = \text{diag}(h_k^1, \ldots, h_k^p)$.

Note that we can derive the updating rules for the entropy-based method by setting $m = 1$.

The membership values are given as

$$u_{ki} = \left[\sum_{j=1}^{c} \left(\frac{D(x_k, P_i)}{D(x_k, P_j)} \right)^{\frac{1}{m-1}} \right]^{-1}, \tag{9.18}$$

for $J_{\text{fcv-pds}}(U, V, F, A)$ and

$$u_{ki} = \frac{\exp\left(-\dfrac{D(x_k, P_i)}{\nu} \right)}{\displaystyle\sum_{j=1}^{c} \exp\left(-\dfrac{D(x_k, P_j)}{\nu} \right)}, \tag{9.19}$$

for $J_{\text{efcv-pds}}(U, V, F, A)$ where

$$D(x_k, P_i) = \sum_{j=1}^{p} h_k^j \left(x_k^j - \sum_{\ell=1}^{q} f_{ik}^{\ell} a_{i\ell}^j - v_i^j \right)^2. \tag{9.20}$$

9.1.3 Linear Fuzzy Clustering with Optimal Completion Strategy

The optimal completion strategy can also be applied to linear fuzzy clustering. In the strategy, missing values are regarded as unknown parameters to be estimated and are completed in the iterative optimization algorithm. In **FCM**, missing values are estimated as the membership weighted means of cluster center values. On the other hand, in **FCV**, we cannot derive such a simple completion rule because the clustering criterion of **FCV** is not composed of component-wise formulation.

In order to estimate missing elements of sample k so that the objective function is minimized, we rewrite the elements of sample k without loss of generality as

$$x_k = \begin{pmatrix} x_k^O \\ x_k^M \end{pmatrix}, \tag{9.21}$$

where x_k^O is a vector having the observed elements of x_k while x_k^M consists of missing ones. In the same way, basis vectors $a_{i\ell}$ and cluster centers v_i can also be decomposed into two vectors. Then, the objective function of **FCV** is given as

$$J_{\text{fcv-ocs}}(U, V, A, X^M) = \sum_{i=1}^{c}\sum_{k=1}^{N}(u_{ki})^m\Bigg\{\|x_k^O - v_i^O\|^2 + \|x_k^M - v_i^M\|^2 \\ - \sum_{\ell=1}^{q}\left\{(a_{i\ell}^O)^\top(x_k^O - v_i^O) + (a_{i\ell}^M)^\top(x_k^M - v_i^M)\right\}^2\Bigg\}. \tag{9.22}$$

For the entropy-based method, it is given as follows:

$$J_{\text{efcv-ocs}}(U, V, A, X^M) = \sum_{i=1}^{c}\sum_{k=1}^{N}u_{ki}\Bigg\{\|x_k^O - v_i^O\|^2 + \|x_k^M - v_i^M\|^2 \\ - \sum_{\ell=1}^{q}\left\{(a_{i\ell}^O)^\top(x_k^O - v_i^O) + (a_{i\ell}^M)^\top(x_k^M - v_i^M)\right\}^2\Bigg\} \\ + \nu\sum_{i=1}^{c}\sum_{k=1}^{N}u_{ki}\log u_{ki}, \tag{9.23}$$

where X^M is the set of missing elements. We can use either $J(U, V, A, X^M) = J_{\text{fcv-ocs}}(U, V, A, X^M)$ or $J(U, V, A, X^M) = J_{\text{efcv-ocs}}(U, V, A, X^M)$.

The iterative algorithm of **FCV** for incomplete data with optimal completion strategy is represented as follows:

Algorithm FCV-OCS: *Fuzzy c-Varieties with Optimal Completion Strategy*

FCV-OCS1. [Generate initial value:] Generate initial values for $\bar{V} = (\bar{v}, \dots, \bar{v}_c)$ and $\bar{A} = (\bar{\mathsf{A}}_1, \dots, \bar{\mathsf{A}}_c)$, and initial values for $\bar{X}^M$.

FCV-OCS2. [Find optimal U:] Calculate

$$\bar{U} = \arg\min_{U \in \mathcal{U}_f} J(U, \bar{V}, \bar{A}, \bar{X}^M). \tag{9.24}$$

FCV-OCS3. [Find optimal V:] Calculate

$$\bar{V} = \arg\min_{V} J(\bar{U}, V, \bar{A}, \bar{X}^M). \tag{9.25}$$

FCV-OCS4. [Find optimal A:] Calculate

$$\bar{A} = \arg\min_{A \in \mathcal{A}} J(\bar{U}, \bar{V}, A, \bar{X}^M), \tag{9.26}$$

where $\bar{A}$ is given from

$$\mathcal{A} = \{\ A = (a_{11}, \ldots, a_{cq})\ :\ ||a_{i\ell}|| = 1\ ;\ a_{i\ell}^\top a_{i\ell'} = 0,\ \ell \neq \ell'\ \}. \tag{9.27}$$

FCV-OCS5. [Find optimal X^M:] Calculate

$$\bar{X}^M = \arg\min_{X^M} J(\bar{U}, \bar{V}, \bar{A}, X^M). \tag{9.28}$$

FCV-OCS6. If the solution $U = (u_{ki})$ is convergent, stop. Otherwise, return to FCV-OCS2.

End FCV-OCS.

This algorithm is equivalent to that of **FCV** except for the additional step of FCV-OCS5.

Considering the condition $\partial L_{\text{fcv-ocs}}/\partial x_k^M = 0$, the optimal estimate of x_k^M composed of the missing elements of sample k is given as

$$\begin{aligned} x_k^M = &\left\{ \sum_{i=1}^{c} (u_{ki})^m \left(I - \sum_{\ell=1}^{q} a_{i\ell}^M (a_{i\ell}^M)^\top \right) \right\}^{-1} \\ &\times \left\{ \sum_{i=1}^{c} (u_{ki})^m \left(\sum_{\ell=1}^{q} a_{i\ell}^M (a_{i\ell}^O)^\top (x_k^O - v_i^O) \right.\right. \\ &\left.\left. + \left(I - \sum_{\ell=1}^{q} a_{i\ell}^M (a_{i\ell}^M)^\top \right) v_i^M \right) \right\}. \end{aligned} \tag{9.29}$$

Note that 0 and I are the zero vector and unit matrix with a certain dimension or rank, respectively. For the entropy-based method, the optimal estimate is given by setting $m = 1$.

9.1.4 Linear Fuzzy Clustering with Nearest Prototype Strategy

While the optimal completion strategy considers the optimal values that minimize the total objective function, it is often the case that the estimated values are far from all the prototypes when the samples including missing elements are shared by several clusters because they are the weighted averages of estimates calculated in clusters. Then, the nearest prototype strategy considers the use of single optimal estimate that is calculated in the cluster having the maximum membership.

Assume that we have the optimal estimate x_{ki}^M in cluster i as

$$\begin{aligned} x_{ki}^M = &\left\{ \left(I - \sum_{\ell=1}^{q} a_{i\ell}^M (a_{i\ell}^M)^\top \right) \right\}^{-1} \\ &\times \left\{ \sum_{\ell=1}^{q} a_{i\ell}^M (a_{i\ell}^O)^\top (x_k^O - v_i^O) + \left(I - \sum_{\ell=1}^{q} a_{i\ell}^M (a_{i\ell}^M)^\top \right) v_i^M \right\}. \end{aligned} \tag{9.30}$$

Using the estimates, we can calculate the clustering criterion $D(x_k, P_i)$ as the distance between sample k and cluster i. The feasible value to be imputed to x_k^M is given as

$$x_k^M = x_{k\phi}^M, \tag{9.31}$$

where

$$D(x_k, P_\phi) = \min\{D(x_k, P_1), \dots, D(x_k, P_c)\}. \tag{9.32}$$

9.1.5 A Comparative Experiment

Let us show a result of comparative experiment on the four strategies. The data set shown in Fig. 9.1 includes 24 samples in 3D space forming two linear clusters and we compared the frequency of the proper clustering results in 500 trials with various initialization sets. The model parameters were set as $c = 2$, $q = 1$ and $m = 2$. When we have a small number of missing elements, all the four strategies can give proper results in many trials. On the other hand, if many samples include missing elements, PDS and OCS are more sensitive to initial partition than WDS and NPS. However, if almost every sample includes missing elements, we cannot apply WDS, and PDS or NPS seems to be a feasible choice.

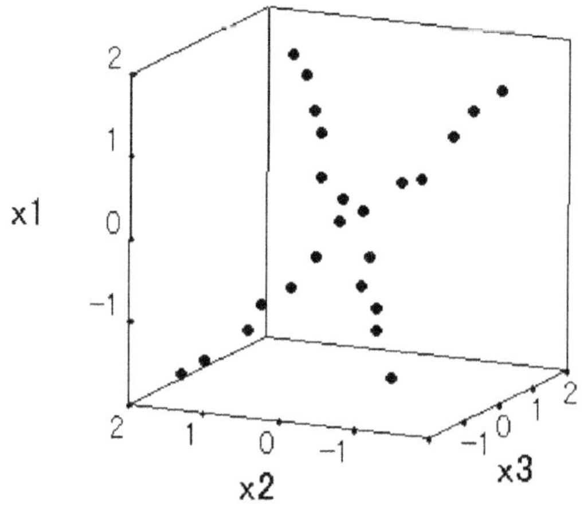

Fig. 9.1. An example of two linear clusters

Table 9.1. Comparison of frequency of proper results (%)

# missing elements	WDS	PDS	OCS	NPS
0	85.0	88.0	85.0	85.0
8	85.6	85.8	91.2	91.6
16	85.6	76.8	78.8	90.6
24	N/A	57.6	46.0	59.8

9.2 Component-wise Robust Clustering

Linear models based on the least squares technique are sensitive to outliers and the derived models are easily influenced by noise. In real applications, two different types of noise often exist. One occurs when the data set includes noise samples and all elements of the noise samples need to be eliminated. The other is data including intra-sample outliers. In this section, we describe a technique for making the linear clustering algorithm robust by handling intra-sample noise and predicting missing values using robust local linear models. Introducing the M-estimation technique, lower rank approximation is performed by ignoring only noise elements. The clustering algorithm is based on iteratively reweighted least squares (IRLS) [53], in which additional weight parameters enable robust approximation to be obtained by using an iterative procedure without solving a non-linear system of equations. Using IRLS approach, the algorithm is reduced to a simple iterative procedure, which is similar to that of the linear fuzzy clustering with the partial distance strategy (PDS). The close connection contributes to applying the method to missing value estimation with robust local linear models.

9.2.1 Robust Principal Component Analysis

When we deal with a data matrix including noise elements, local models based on least squares techniques are easily distorted. In the analysis of large scale databases with high-dimensional observations, in many cases, almost every sample includes a few noise elements and conventional robust clustering methods such as possibilistic clustering [87] or noise clustering [22, 23] fail to derive good results because all noise samples are eliminated, even though only a few elements of the samples are noise.

The M-estimation technique is a useful method for estimating robust models [46, 64]. The goal of the technique is to derive a solution ignoring outliers that do not conform to the assumed statistical model.

For robust principal component analysis of a noisy data set including intra-sample outliers, de la Torre *et al.* [24] proposed a robust subspace learning technique based on robust M-estimation. In the robust PCA technique, the energy function to be minimized is defined as

$$J_{\mathrm{rpca}}(V, F, A) = \sum_{k=1}^{N} \sum_{j=1}^{p} \rho \left(x_k^j - \sum_{\ell=1}^{q} f_k^\ell a_\ell^j - v^j \right), \tag{9.33}$$

where $\rho(\cdot)$ is a class of robust ρ-functions [53]. In [24], the Geman-McClure error function [40] was used,

$$\rho(x) = \frac{x^2}{x^2 + \sigma_j^2}, \tag{9.34}$$

where σ_j is a scale parameter that controls the convexity of the robust function. In the optimization process, the value of σ_j is decreased by the deterministic annealing technique. To solve the minimization problem, the use of both the iteratively reweighted least-squares (IRLS) technique [53] and the gradient descent method with a local quadratic approximation was proposed.

9.2.2 Robust Local Principal Component Analysis

Next, we introduce a technique for handling intra-sample outliers in local PCA based on component-wise least squares approximation [56]. In order to robustify the local PCA model, we apply the ρ-function to the **FCV** algorithm using the least squares criterion of (8.53). The objective function of robust **FCV** is defined as follows:

$$J_{\mathrm{rfcv}}(U,V,F,A)=\sum_{i=1}^{c}\sum_{k=1}^{N}(u_{ki})^m\sum_{j=1}^{p}\rho\left(x_k^j-\sum_{\ell=1}^{q}f_{ik}^{\ell}a_{i\ell}^{j}-v_i^j\right). \quad (9.35)$$

When we use the entropy-based method, the objective function is defined as

$$J_{\mathrm{erfcv}}(U,V,F,A)=\sum_{i=1}^{c}\sum_{k=1}^{N}u_{ki}\sum_{j=1}^{p}\rho\left(x_k^j-\sum_{\ell=1}^{q}f_{ik}^{\ell}a_{i\ell}^{j}-v_i^j\right)$$
$$+\nu\sum_{i=1}^{c}\sum_{k=1}^{N}u_{ki}\log u_{ki}. \quad (9.36)$$

To obtain a unique solution, the objective function is minimized under the constraints (9.6)-(9.8) and $\mathsf{A}_i^\top\mathsf{A}_i$ is diagonal. For the entropy-based method, we put $m=1$.

The optimal solution cannot be derived by the conventional iterative procedure because the clustering criterion is transformed by the non-linear ρ function. Thus, the solution is derived based on the IRLS technique, in which the minimization problem is formulated as a weighted least squares problem with an $N\times p$ weight matrix $\mathsf{W}_i=(w_{ik}^j)$ in each cluster. w_{ik}^j represents the positive weight for previous residual $e_{ik}^j=x_k^j-\sum_{\ell=1}^{q}f_{ik}^{\ell}a_{i\ell}^j-v_i^j$. For the Geman-McClure ρ function, the weight w_{ik}^j is given by

$$w_{ik}^j=\frac{\psi(e_{ik}^j,\sigma_j)}{e_{ik}^j}, \quad (9.37)$$

where

$$\psi(e_{ik}^j,\sigma_j)=\frac{\partial\rho(e_{ik}^j)}{\partial e_{ik}^j}$$
$$=\frac{2e_{ik}^j\sigma_j^2}{\left((e_{ik}^j)^2+\sigma_j^2\right)^2}, \quad (9.38)$$

and the objective function is modified as follows:

$$J_{\mathrm{rfcv'}}(U,V,F,A)=\sum_{i=1}^{c}\sum_{k=1}^{N}(u_{ki})^m\sum_{j=1}^{p}w_{ik}^j\left(x_k^j-\sum_{\ell=1}^{q}f_{ik}^{\ell}a_{i\ell}^{j}-v_i^j\right)^2. \quad (9.39)$$

For the entropy-based method,

$$J_{\mathrm{erfcv'}}(U,V,F,A) = \sum_{i=1}^{c}\sum_{k=1}^{N} u_{ki} \sum_{j=1}^{p} w_{ik}^{j} \left(x_k^j - \sum_{\ell=1}^{q} f_{ik}^{\ell} a_{i\ell}^{j} - v_i^j \right)^2$$
$$+\nu \sum_{i=1}^{c}\sum_{k=1}^{N} u_{ki} \log u_{ki}. \tag{9.40}$$

We can use either $J(U,V,F,A) = J_{\mathrm{rfcv'}}(U,V,F,A)$ or $J(U,V,F,A) = J_{\mathrm{erfcv'}}(U,V,F,A)$. Minimization of the modified objective function $J(U,V,F,A)$ approximately achieves the optimization of (9.35) or (9.36) because the first derivative with fixed w_{ik}^j is similar to that of the original objective functions as follows:

$$\frac{\partial J_{\mathrm{rfcv'}}}{\partial e_{ik}^j} = 2(u_{ki})^m w_{ik}^j e_{ik}^j$$
$$\cong \frac{2(u_{ki})^m \sigma_j^2 e_{ik}^j}{\left((e_{ik}^j)^2 + \sigma_j^2\right)^2} = \frac{\partial J_{\mathrm{rfcv}}}{\partial e_{ik}^j}, \tag{9.41}$$

and it is also the case for the entropy-based method with $m = 1$. If the parameter σ_j has a large value, the modified objective function gives similar results to the **FCV** algorithm since all the weights w_{ik}^j have similar values.

The algorithm for the robust FCV clustering can be written as follows:

Algorithm Robust FCV: *Robust Fuzzy c-Varieties* [56]

Robust FCV1. [Generate initial value:] Generate initial values for $\bar{U} = (\bar{\mathsf{U}}_1, \ldots, \bar{\mathsf{U}}_c)$, $\bar{V} = (\bar{v}_1, \ldots, \bar{v}_c)$, $\bar{F} = (\bar{\mathsf{F}}_1, \ldots, \bar{\mathsf{F}}_c)$ and $\bar{A} = (\bar{\mathsf{A}}_1, \ldots, \bar{\mathsf{A}}_c)$, and normalize them so that they satisfy the constraints (9.6)-(9.8) and $\bar{\mathsf{A}}_i^\top \bar{\mathsf{A}}_i$ is diagonal.

Robust FCV2. [Calculate initial W:] Calculate initial values for the set of weight parameters $W = (\mathsf{W}_1, \ldots, \mathsf{W}_c)$.

Robust FCV3. [Find optimal U:] Calculate

$$\bar{U} = \arg\min_{U \in \mathcal{U}_f} J(U, \bar{V}, \bar{F}, \bar{A}). \tag{9.42}$$

Robust FCV4. [Find optimal V:] Calculate

$$\bar{V} = \arg\min_{V} J(\bar{U}, V, \bar{F}, \bar{A}). \tag{9.43}$$

Robust FCV5. [Find optimal F:] Calculate

$$\bar{F} = \arg\min_{F} J(\bar{U}, \bar{V}, F, \bar{A}), \tag{9.44}$$

and normalize them so that they satisfy the constraints (9.6) and (9.7).

Robust FCV6. [Find optimal A:] Calculate

$$\bar{A} = \arg\min_{A} J(\bar{U}, \bar{V}, \bar{F}, A), \tag{9.45}$$

and transform them so that each $\mathsf{A}_i^\top \mathsf{A}_i$ is diagonal.

Robust FCV7. [Test convergence:] If all parameters except for weight parameters W are convergent, go to Robust FCV8; else go to Robust FCV3.
Robust FCV8. [Test convergence of W:] Calculate W. If the weight parameters are convergent, stop; else go to Robust FCV3.
End Robust FCV.

The orthogonal matrices in Robust FCV1, Robust FCV5 and Robust FCV6 are obtained by such a technique as Gram-Schmidt's orthgonalization.

Usually the initial value of the parameter σ_j is large and is annealed in the optimization process. The initial partitioning is then given by the **FCV** algorithm.

To derive the optimal A and V, we rewrite the objective function (9.39) as follows:

$$J_{\mathrm{rfcv}'} = \sum_{i=1}^{c}\sum_{j=1}^{p}(\tilde{x}_j - \mathsf{F}_i\tilde{a}_i^j - 1_N v_i^j)^\top \mathsf{U}_i^m \tilde{\mathsf{W}}_{ij}(\tilde{x}_j - \mathsf{F}_i\tilde{a}_i^j - 1_N v_i^j), \quad (9.46)$$

where

$$\mathsf{X} = (\tilde{x}_1, \ldots, \tilde{x}_p),$$
$$\mathsf{A}_i = (\tilde{a}_i^1, \ldots, \tilde{a}_i^p)^\top,$$
$$\tilde{\mathsf{W}}_{ij} = \mathrm{diag}(w_{i1}^j, \ldots, w_{iN}^j).$$

From $\partial J_{\mathrm{rfcv}'}/\partial\tilde{a}_i^j = 0_q$ and $\partial J_{\mathrm{rfcv}'}/\partial v_i^j = 0$, we have

$$\tilde{a}_i^j = (\mathsf{F}_i^\top \mathsf{U}_i^m \tilde{\mathsf{W}}_{ij}\mathsf{F}_i)^{-1}\mathsf{F}_i^\top \mathsf{U}_i^m \tilde{\mathsf{W}}_{ij}(\tilde{x}_j - 1_N v_i^j), \quad (9.47)$$
$$v_i^j = (1_N^\top \mathsf{U}_i^m \tilde{\mathsf{W}}_{ij} 1_N)^{-1} 1_N^\top \mathsf{U}_i^m \tilde{\mathsf{W}}_{ij}(\tilde{x}_j - \mathsf{F}_i\tilde{a}_i^j). \quad (9.48)$$

In the same way, we can derive the optimal F. The objective function (9.39) is equivalent to

$$J_{\mathrm{rfcv}'} = \sum_{i=1}^{c}\sum_{k=1}^{N}(u_{ki})^m (x_k - \mathsf{A}_i f_{ik} - v_i)^\top \mathsf{W}_{ik}(x_k - \mathsf{A}_i f_{ik} - v_i), \quad (9.49)$$

and $\partial J_{\mathrm{rfcv}'}/\partial f_{ik} = 0_q$ yields

$$f_{ik} = (\mathsf{A}_i^\top \mathsf{W}_{ik}\mathsf{A}_i)^{-1}\mathsf{A}_i^\top \mathsf{W}_{ik}(x_k - v_i), \quad (9.50)$$

where $\mathsf{W}_{ik} = \mathrm{diag}(w_{ik}^1, \ldots, w_{ik}^p)$.

Note that we can derive the updating rules for the entropy-based method by setting $m = 1$.

The membership values are given as

$$u_{ki} = \left[\sum_{j=1}^{c}\left(\frac{D(x_k, P_i)}{D(x_k, P_j)}\right)^{\frac{1}{m-1}}\right]^{-1}, \quad (9.51)$$

for $J_{\mathrm{rfcv}'}(U,V,F,A)$ and

$$u_{ki}=\frac{\exp\left(-\dfrac{D(x_k,P_i)}{\nu}\right)}{\displaystyle\sum_{j=1}^{c}\exp\left(-\dfrac{D(x_k,P_j)}{\nu}\right)}, \tag{9.52}$$

for $J_{\mathrm{erfcv}'}(U,V,F,A)$ where

$$D(x_k,P_i)=\sum_{j=1}^{p}w_{ik}^{j}\left(x_k^j-\sum_{\ell=1}^{q}f_{ik}^{\ell}a_{i\ell}^{j}-v_i^j\right)^2. \tag{9.53}$$

9.2.3 Handling Missing Values and Application to Missing Value Estimation

Aside from noise or outliers, missing values are also common in real world data analysis. In such cases, a priori knowledge about observations is often available although many noise elements are not identified prior to analysis.

The robust **FCV** algorithm is also applied to incomplete data sets. To handle missing values, weight parameter w_{ik}^j is redefined as follows:

$$w_{ik}^{j}=\begin{cases}\dfrac{2\sigma_j^2}{((e_{ik}^j)^2+\sigma_j^2)^2} & ; x_k^j \text{ is observed.}\\ 0 & ; x_k^j \text{ is missing.}\end{cases} \tag{9.54}$$

If all the weights w_{ik}^j for observed elements x_k^j are 1, the robust FCV method is equivalent to the **FCV** algorithm with partial distance strategy (**FCV-PDS**).

Once local linear models are estimated, missing values of data matrix X can be predicted using the corresponding elements of the approximation matrix Y_i because Y_i includes no missing values. The cluster of a sample is determined based on the maximum membership rule and a missing value x_k^j is predicted from the corresponding value y_{ik}^j. This means that missing values are estimated based on the assumption that data points including missing values should exist on the nearest points to the prototypical linear varieties spanned by local principal component vectors. Here, it should be noted that this imputation strategy is more similar to nearest prototype strategy (NPS) than optimal completion strategy (OCS). The lower rank approximation of each data point x_k is, however, estimated in each cluster and the clustering criterion is calculated for each cluster using the corresponding y_{ik}. In this sense, the missing value estimation approach using PDS is a hybrid version of OCS and NPS.

9.2.4 An Illustrative Example

An numerical experiment was performed using an artificial data set [56]. The 3-D data set composed of 24 samples forms two lines, with the goal of the

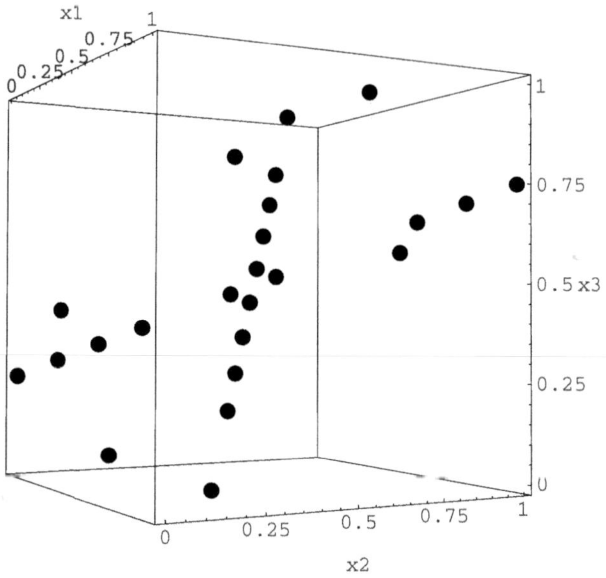

Fig. 9.2. An example of noisy data set

analysis being to capture these lines. Replacing randomly selected elements with noise values, a noisy data set including 21% noise samples was constructed. Figures 9.2 and 9.3 show 3-D plot and 2-D projections of the noisy data set. Note that each "noise sample" includes only one noise element, i.e., the noise sample points are on the line in one of three projections in Fig. 9.3. Figures 9.4 and 9.5 show clustering results derived by the standard **FCV** algorithm and the robust **FCV** algorithm, respectively. In the figures, the two lines represent the prototypes of two clusters (∘ and •). The scale parameter σ_j was annealed by the following schedule:

$$\sigma_j^2 = \frac{0.5}{\log(t+2)}, \tag{9.55}$$

where t is the iteration index. Although the prototypes of the standard **FCV** algorithm were influenced by outliers, the robust **FCV** algorithm was able to capture the two lines properly. Note that the robust **FCV** algorithm ignored only the noise elements and assigned the noise samples into proper clusters.

9.2.5 A Potential Application: Collaborative Filtering

A potential application is collaborative filtering. Automated collaborative filtering is a useful tool for reducing information overload and has been considerably successful in many areas [86, 132, 143]. Filtering systems are built on the assumption that items to be recommended are the items preferred by users who have similar interests to the active user. The task is then to predict the applicability of items to the active user based on a database of users' ratings. The

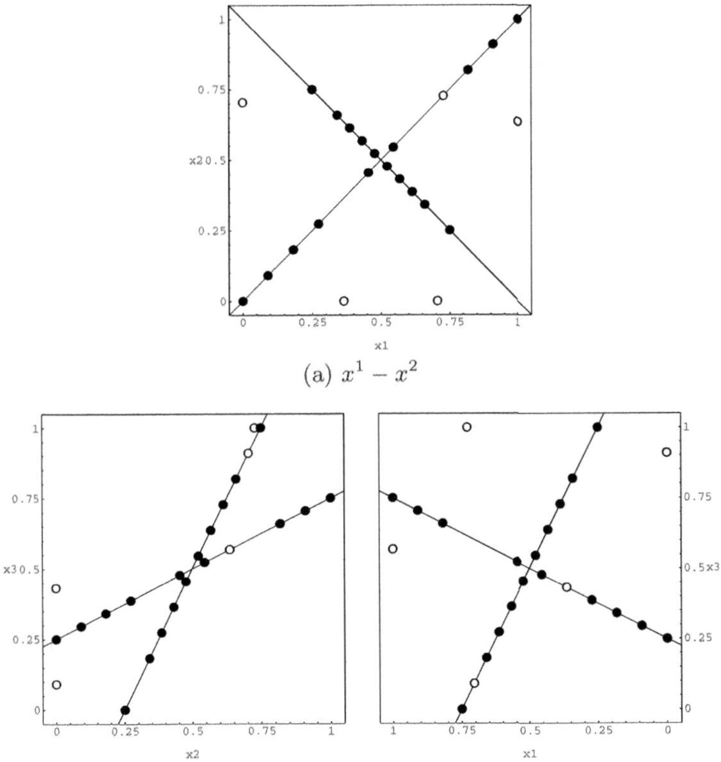

(a) $x^1 - x^2$

(b) $x^2 - x^3$ (c) $x^1 - x^3$

Fig. 9.3. 2-D projections of a noisy data set

problem space can be formulated as a matrix of users versus items X, with its element x_k^j representing the k-th user's rating of a specific item j and the goal is to predict the missing values in the data matrix [51].

Using the estimated local linear models, the ratings of new active users can also be predicted [56, 61]. Memberships and principal component scores of the new active users are estimated by (9.51) (or (9.52)) and (9.50), respectively, and y_{ik}^j is predicted using the following equation:

$$y_{ik}^j = \sum_{\ell=1}^{q} f_{ik}^{\ell} a_{i\ell}^j + v_i^j. \tag{9.56}$$

Even when we have an enormous number of users, missing values can be predicted by using only (9.51) (or (9.52)), (9.50) and local principal component matrices A. In this sense, the prediction method is an efficient technique with fewer memory requirements [61].

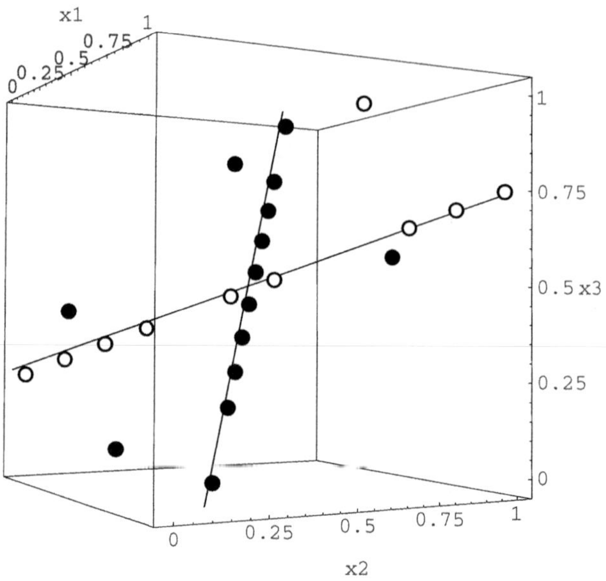

Fig. 9.4. Clustering result by **FCV**

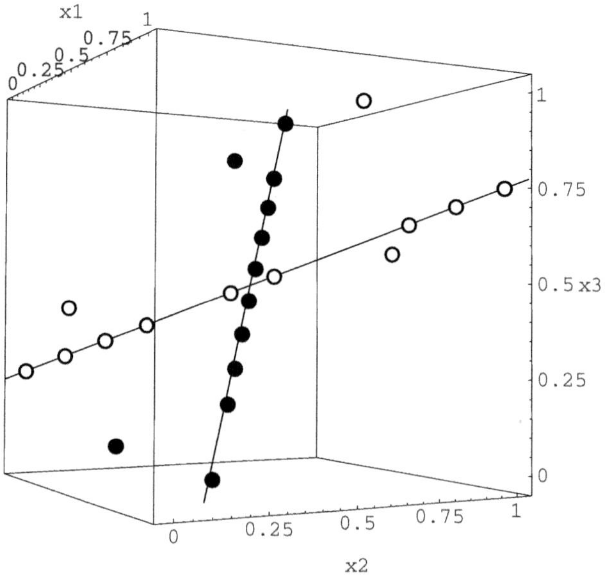

Fig. 9.5. Clustering result by **Robust FCV**

9.3 Local Minor Component Analysis Based on Least Absolute Deviations

In this section, we introduce a modified linear clustering method that is not sensitive to noise samples using absolute deviations. Since the objective function composed of minor component scores presented in Section 8.2.3 is simpler than that of **FCV**, it is easy to apply absolute deviations if prototypes are hyperplanes ($q = p - 1$). By replacing the clustering criterion with the absolute distance, the objective function is defined as follows [62]:

$$\begin{aligned} J_{\text{lmca}}(U, V, A) &= \sum_{i=1}^{c}\sum_{k=1}^{N}(u_{ki})^m D(x_k, P_i) \\ &= \sum_{i=1}^{c}\sum_{k=1}^{N}(u_{ki})^m |a_{mi}^\top (x_k - v_i)| \end{aligned} \tag{9.57}$$

For the entropy-based method,

$$\begin{aligned} J_{\text{elmca}}(U, V, A) &= \sum_{i=1}^{c}\sum_{k=1}^{N} u_{ki} D(x_k, P_i) + \nu \sum_{i=1}^{c}\sum_{k=1}^{N} u_{ki} \log u_{ki} \\ &= \sum_{i=1}^{c}\sum_{k=1}^{N} u_{ki} |a_{mi}^\top (x_k - v_i)| + \nu \sum_{i=1}^{c}\sum_{k=1}^{N} u_{ki} \log u_{ki}. \end{aligned} \tag{9.58}$$

A is the set of the unit normal vector of the q dimensional prototypical hyperplane a_{mi} where $q = p - 1$. $D(x_k, P_i) = |a_{mi}^\top (x_k - v_i)|$ represents the absolute distance between data point k and prototypical hyperplane P_i. The solution algorithm is a fixed-point iteration scheme. In the following, we show how a_{mi} and v_i can be calculated based on least absolute deviations.

9.3.1 Calculation of Optimal Local Minor Component Vectors

Here, we discuss the property of the optimal prototype that minimizes the objective function with fixed U and V.

Proposition 9.3.1. *The calculation of the optimal local minor component vector is reduced to a combinatorial optimization problem because the optimal prototype of each cluster paths through at least $p - 1$ different sample points with fixed U and V. The optimal hyperplane and local minor component vector are derived by searching the combination of $(p - 1)$ different points such that the objective function is minimized.*

First, we consider the situation where the data dimensionality p is 2 and the goal is to derive the optimal unit vector a_{mi} in cluster i. We have the next lemma.

Lemma 9.3.2. *When U and V are fixed, there is at least one data point x_k that satisfies the condition $a_{mi}^\top(x_k - v_i) = 0$ in each cluster, that is, there is at least one data point on the optimal prototype in each cluster.*

Proof. For the illustrative purposes, let us transform the coordinate system from the original 2-D Cartesian coordinate to the spherical polar system in each cluster. In cluster i, data point $x_k = (x_k^1, x_k^2)$ is represented by using the distance from the cluster center $\kappa_{ki} = ||x_k - v_i||$ and the angle ϕ_{ki} as follows:

$$x_k^1 - v_i^1 = \kappa_{ki} \cos \phi_{ki}, \tag{9.59}$$

$$x_k^2 - v_i^2 = \kappa_{ki} \sin \phi_{ki}. \tag{9.60}$$

The minor component vector $a_{mi} = (a_{mi}^1, a_{mi}^2)$ is also noted as

$$a_{mi}^1 = \cos \psi_i, \tag{9.61}$$

$$a_{mi}^2 = \sin \psi_i, \tag{9.62}$$

where ψ_i is the angle that defines the direction of the prototypical line in the 2-D space.

The angles vary as $0 \leq \phi_{ki} \leq 2\pi$, $0 \leq \psi_i \leq 2\pi$. In the coordinate system, the objective function (9.57) is written as follows:

$$J_{\text{lmca}}(U, V, A) = \sum_{i=1}^{c} \sum_{k=1}^{N} (u_{ki})^m \kappa_{ki} \delta_{ki} (\cos \phi_{ki} \cos \psi_i + \sin \phi_{ki} \sin \psi_i), \tag{9.63}$$

where δ_{ki} is a binary variable,

$$\delta_{ki} = \begin{cases} 1 & ; \kappa_{ki}(\cos \phi_{ki} \cos \psi_i + \sin \phi_{ki} \sin \psi_i) \geq 0 \\ -1 & ; \kappa_{ki}(\cos \phi_{ki} \cos \psi_i + \sin \phi_{ki} \sin \psi_i) < 0 \end{cases} \tag{9.64}$$

Then, the minimization problem of finding the optimal a_{mi} under the condition of $a_{mi}^\top a_{mi} = 1$ is equivalent to the problem of finding the optimal angle ψ_i that minimizes the objective function (9.63).

If all δ_{ki} are fixed, i.e., ψ_i can vary under the condition (9.64) for the fixed δ_{ki}, the second derivative of (9.63) is calculated as

$$\begin{aligned} \frac{\partial^2 J_{\text{lmca}}}{\partial (\psi_i)^2} &= -\sum_{k=1}^{N} (u_{ki})^m \kappa_{ki} \delta_{ki} (\cos \phi_{ki} \cos \psi_i + \sin \phi_{ki} \sin \psi_i) \\ &= -J_{\text{lmca}} < 0. \end{aligned} \tag{9.65}$$

Unless all sample points are on the prototypical line, the criterion of J_{lmca} is greater than 0 and the negative second derivative implies that the objective function is convex with respect to ψ_i and the optimal ψ_i is an extremal point where ψ_i is equivalent to at least one ϕ_{ki}. Thus the lemma is proved. □

Next, we generalize Lemma 9.3.2 to p dimensional case.

Lemma 9.3.3. *When U and V are fixed, there are at least $(p-1)$ data points $x_k \in (x_{\tau_1}, \dots, x_{\tau_{p-1}})$ that satisfy the condition $a_{mi}^\top(x_k - v_i) = 0$ in each cluster, that is, there are at least $(p-1)$ data points on the optimal prototype in each cluster.*

Proof. In the same way with Lemma 9.3.2, using the distance κ_{ki} and the angles $\phi_{ki}^1, \dots, \phi_{ki}^{p-1}$, data point k with p dimensional observation in cluster i is represented as

$$\begin{aligned} x_k^1 - v_i^1 &= \kappa_{ki} \cos\phi_{ki}^1, \\ x_k^2 - v_i^2 &= \kappa_{ki} \sin\phi_{ki}^1 \cos\phi_{ki}^2, \\ &\dots \\ x_k^{p-1} - v_i^{p-1} &= \kappa_{ki} \sin\phi_{ki}^1 \dots \sin\phi_{ki}^{p-2} \cos\phi_{ki}^{p-1}, \\ x_k^p - v_i^p &= \kappa_{ki} \sin\phi_{ki}^1 \dots \sin\phi_{ki}^{p-2} \sin\phi_{ki}^{p-1}, \end{aligned}$$

in a spherical polar system. And the unit vector a_{mi} is also written as

$$\begin{aligned} a_{mi}^1 &= \cos\psi_i^1, \\ a_{mi}^2 &= \sin\psi_i^1 \cos\psi_i^2, \\ &\dots \\ a_{mi}^{p-1} &= \sin\psi_i^1 \dots \sin\psi_i^{p-2} \cos\psi_i^{p-1}, \\ a_{mi}^p &= \sin\psi_i^1 \dots \sin\psi_i^{p-2} \sin\psi_i^{p-1}, \end{aligned}$$

where $\psi_i^1, \dots, \psi_i^{p-1}$ are angles. The angles satisfy $0 \le \phi_{ki}^1, \dots, \phi_{ki}^{p-2} \le \pi$, $0 \le \psi_i^1, \dots, \psi_i^{p-2} \le \pi$ and $0 \le \phi_{ki}^{p-1} \le 2\pi$, $0 \le \psi_i^{p-1} \le 2\pi$. The objective function is also defined in the similar form to (9.63) and the second derivative with fixed δ_{ki} is calculated as

$$\frac{\partial^2 J_{\text{lmca}}}{\partial(\psi_i^1)^2} = -J_{\text{lmca}} < 0. \tag{9.66}$$

So, the hessian matrix is not positive semi-definite. If the objective function has a minimum at a certain point where the gradient is 0, the hessian matrix is positive semi-definite. Therefore, we have no minimum points in the searching area and the optimal minor component vector is given at an extremal point where at least one data point is on the prototypical hyperplane.

Consequently, the objective function is optimized considering that a certain sample data point is on the prototype. Suppose that data point τ_1 is on the prototype and the coordinate system is rotated so that $\phi_{\tau_1 i}^1 = \dots = \phi_{\tau_1 i}^{p-1} = 0$, i.e., the vector $(x_{\tau_1} - v_i)$ is the first basis vector in the coordinate system. The optimization problem is reduced to the problem of finding the optimal angles $\psi_i^2, \dots, \psi_i^{p-1}$ in the $p-1$ dimensional subspace whose normal vector is $(x_{\tau_1} - v_i)$ under the condition that all data points except for τ_1 are projected on the subspace and $a_{mi}^1 = 0$, i.e., $\psi_i^1 = \pi/2$. The reduced problem also has above property in the $p-1$ dimensional data space in which ψ_i^1 is replaced with ψ_i^2. The similar discussion is iterated $p-1$ times. Thus the lemma is proved. □

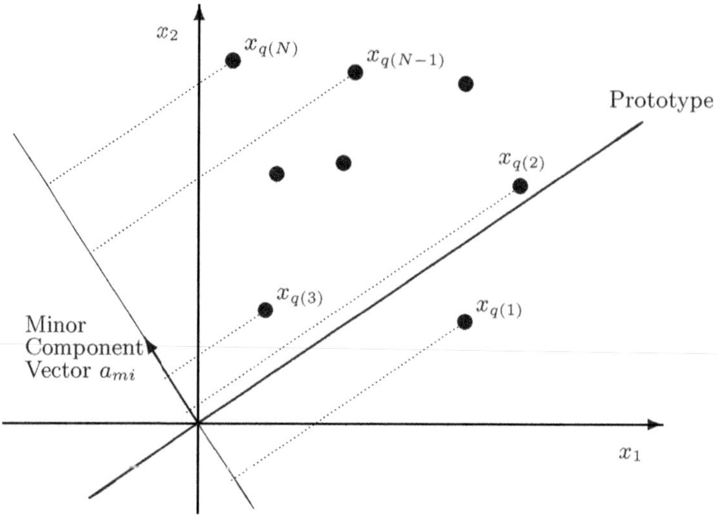

Fig. 9.6. Ordering of sample data points

From Lemma 9.3.3, we can see that the optimal hyperplane is spanned by the cluster center and $p-1$ linearly independent sample data points. Note that we also have same property for the entropy-based method.

9.3.2 Calculation of Optimal Cluster Centers

Next, we consider the property of the optimal cluster centers. With fixed U and A, the objective function to be minimized is regarded as the weighted sum of L_1 errors in the 1-D space spanned by a_{mi}.

In the case of the FCM clustering, the L_1 metric based methods have been studied by several researchers [12, 73] and Miyamoto and Agusta [101] proposed to use a linear search algorithm for calculating cluster centers in an iteration procedure. The algorithm is based on the property that v_i^j is equivalent to at least one x_k^j.

In the same way, we can use the property that at least one x_k is on the optimal prototype. Assume that $x_1, \ldots, x_N$ are sorted and their subscripts are changed using a permutation function $q(k), k = 1, \ldots, N$, as shown in Fig. 9.6, where

$$a_{mi}^\top x_{q(1)} \leq a_{mi}^\top x_{q(2)} \leq \cdots \leq a_{mi}^\top x_{q(N)}. \tag{9.67}$$

The optimization problem is reduced to finding a weighted median. Here, the cluster center has the arbitrariness like that of the **FCV** algorithm, i.e., absolute deviations are invariant so long as v_i is on the optimal prototypical hyperplane. Because L_2 norm is invariant with respect to an orthonormal basis, the **FCV** algorithm adopted the weighted average as the cluster center [63]. However, L_1 norm is not invariant with respect to an orthonormal basis. And the median of (9.67) is not necessarily a good cluster center because such an inappropriate noise sample data as $x_{q(3)}$ of Fig. 9.6 might be chosen. Then, we use the following algorithm, which derives a unique solution in an orthonormal system where a_{mi} is the first basis vector.

begin
  $v_i := 0_p$;
  ORTHONORMALIZING(a_{mi}, B);
  $j := 0$;
  while $(j < p)$ **do begin**
    $j := j + 1$;
    ORDERING(β_j, X);
    $S := -\sum_{k=1}^{N} (u_{ki})^m$;
    $r := 0$;
    while $(S < 0)$ **do begin**
      $r := r + 1$;
      $S := S + 2(u_{q(r)i})^m$;
    end;
    $v_i := v_i + (\beta_j^\top x_{q(r)})\beta_j$;
  end;
  output v_i;
end.

In the algorithm, ORTHONORMALIZING(a_{mi}, B) denotes the subroutine that derives a set of orthonormal basis vectors $B = (\beta_1, \ldots, \beta_p)$ by the Gram-Schmidt's orthogonalization where $\beta_1 = a_{mi}$. ORDERING(β_j, X) performs the ordering of the subscripts $q(k)$ so that

$$\beta_j^\top x_{q(1)} \leq \beta_j^\top x_{q(2)} \leq \cdots \leq \beta_j^\top x_{q(N)}. \tag{9.68}$$

Although the ordering of sample data points is necessary in each iteration, calculation of cluster centers is achieved by the simple and fast linear search algorithm.

9.3.3 An Illustrative Example

Let us show a comparative result between the **FCV** algorithm and local minor component analysis (LMCA) based on least absolute deviations. An artificial data set shown in Fig. 9.7 is composed of 100 samples distributed on a 2-D space forming two lines buried in 50 noise data points. We partitioned the data set into two clusters with $m = 2$. Figures 9.8 and 9.9 show the clustering results in which the sample points are crisply classified into two clusters ($\square$ and $\times$) based on the maximum membership rule and their prototypical lines that pass through cluster centers ($\bullet$) are depicted by dotted lines. The short lines in the result of **LMCA** represent the directions of the minor component vectors. Figure 9.8 shows that results of the **FCV** algorithm are influenced by noise data and we cannot capture

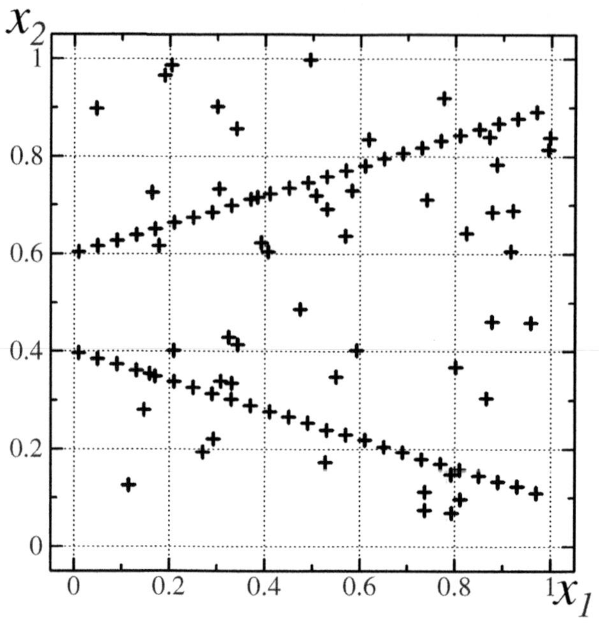

Fig. 9.7. An example of noisy data set

local linear structures properly because the algorithm uses squared distances as the clustering criterion. On the other hand, the prototypes derived by **LMCA** based on least absolute deviations represented the local linear structures properly using absolute deviations. From the above results, the robust algorithm is shown to be useful for capturing local linear structures when majority of samples are on prototypical hyperplanes.

9.4 Local PCA with External Criteria

In real applications, it is often the case that we fail to derive practical knowledge from intrinsic models of databases when we use feature values extracted by linear models because they are influenced by external variables. Yanai [174] proposed PCA with external criteria that extracts latent variables uncorrelated to some external criteria. In the technique, influences of external criteria are first removed from a data matrix by using regression analysis and latent variables are estimated from remaining components. In this section, an enhanced PCA technique for local linear modeling is introduced by using fuzzy c-regression models (FCRM) [48] instead of regression analysis.

9.4.1 Principal Components Uncorrelated with External Criteria

Assume that r observed variables $y = (y^1, \ldots, y^r)^\top$ are influenced by p external criteria $x = (x^1, \ldots, x^p)^\top$ and the two data matrices composed of N samples are given as follows:

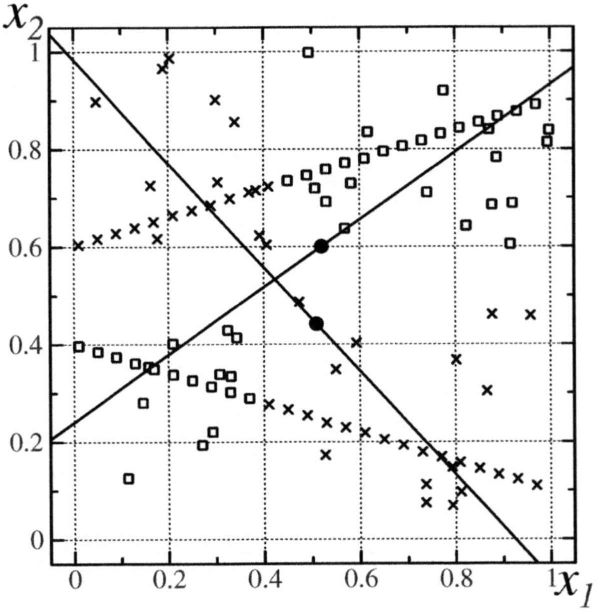

Fig. 9.8. Result by standard **FCV**

$$\mathsf{X} = \begin{pmatrix} x_1^\top \\ x_2^\top \\ \vdots \\ x_N^\top \end{pmatrix} = \begin{pmatrix} x_1^1 & x_1^2 & \cdots & x_1^p \\ x_2^1 & x_2^2 & \cdots & x_2^p \\ \vdots & \vdots & \ddots & \vdots \\ x_N^1 & x_N^2 & \cdots & x_N^p \end{pmatrix},$$

$$\mathsf{Y} = \begin{pmatrix} y_1^\top \\ y_2^\top \\ \vdots \\ y_N^\top \end{pmatrix} = \begin{pmatrix} y_1^1 & y_1^2 & \cdots & y_1^r \\ y_2^1 & y_2^2 & \cdots & y_2^r \\ \vdots & \vdots & \ddots & \vdots \\ y_N^1 & y_N^2 & \cdots & y_N^r \end{pmatrix},$$

where each variable is mean corrected as

$$\sum_{k=1}^{N} x_k^j = 0 \quad ; \quad j = 1, \ldots, p, \tag{9.69}$$

$$\sum_{k=1}^{N} y_k^j = 0 \quad ; \quad j = 1, \ldots, r. \tag{9.70}$$

Here, the goal is to extract principal components of y independent of external criteria $x^1, \ldots, x^p$, i.e., we try to estimate principal components uncorrelated to

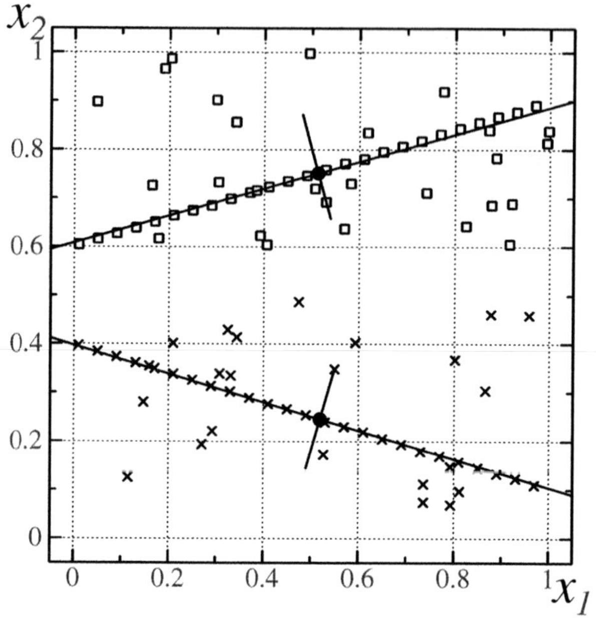

Fig. 9.9. Result by **LMCA** based on least absolute deviations

the external criteria. Yanai [174] proposed a technique for estimating principal components that are independent of some external criteria. In the technique, influences of external criteria are first removed from observed variables by considering the following linear regression model:

$$y^\top = x^\top \mathsf{B} + e^\top, \tag{9.71}$$

where B is a partial regression coefficients matrix,

$$\mathsf{B} = \begin{pmatrix} \beta_1^1 & \beta_2^1 & \dots & \beta_r^1 \\ \beta_1^2 & \beta_2^2 & \dots & \beta_r^2 \\ \vdots & \vdots & \ddots & \vdots \\ \beta_1^p & \beta_2^p & \dots & \beta_r^p \end{pmatrix}, \tag{9.72}$$

and e is an error term. Minimizing the least square criterion

$$\begin{aligned} J_{\mathrm{ra}}(\mathsf{B}) &= \sum_{k=1}^{N} \|y_k^\top - x_k^\top \mathsf{B}\|^2 \\ &= \mathrm{tr}\Big((\mathsf{Y} - \mathsf{XB})^\top (\mathsf{Y} - \mathsf{XB})\Big), \end{aligned} \tag{9.73}$$

the optimal B is estimated as

$$\mathsf{B} = (\mathsf{X}^\top \mathsf{X})^{-1} \mathsf{X}^\top \mathsf{Y}, \tag{9.74}$$

and data matrix Y is decomposed into

$$\begin{aligned}\mathsf{Y} &= \mathsf{X}(\mathsf{X}^\top\mathsf{X})^{-1}\mathsf{X}^\top\mathsf{Y} + \left(I_N - \mathsf{X}(\mathsf{X}^\top\mathsf{X})^{-1}\mathsf{X}^\top\right)\mathsf{Y}\\ &= \mathsf{Y}_X + \mathsf{Y}_{\bar{X}},\end{aligned} \tag{9.75}$$

where the first term of (9.75) is the matrix composed of the elements predictable perfectly by external criteria and the second term is composed of the elements independent of X. So covariance matrix C_{YY} is also decomposed into

$$\begin{aligned}C_{YY} &= \frac{1}{N}\mathsf{Y}^\top\mathsf{Y}\\ &= \frac{1}{N}\mathsf{Y}^\top(\mathsf{Y}_X + \mathsf{Y}_{\bar{X}})\\ &= \frac{1}{N}\mathsf{Y}^\top\mathsf{X}(\mathsf{X}^\top\mathsf{X})^{-1}\mathsf{X}^\top\mathsf{Y} + \frac{1}{N}\left(\mathsf{Y}^\top\mathsf{Y} - \mathsf{Y}^\top\mathsf{X}(\mathsf{X}^\top\mathsf{X})^{-1}\mathsf{X}^\top\mathsf{Y}\right)\\ &= C_{YX}C_{XX}^{-1}C_{XY} + (C_{YY} - C_{YX}C_{XX}^{-1}C_{XY})\\ &= C_{YY.X} + C_{YY.\bar{X}},\end{aligned} \tag{9.76}$$

$$C_{YY.X} = C_{YX}C_{XX}^{-1}C_{XY}, \tag{9.77}$$

$$C_{YY.\bar{X}} = C_{YY} - C_{YX}C_{XX}^{-1}C_{XY}, \tag{9.78}$$

where C_{XX} is the variance-covariance matrix of X, and C_{XY} is the covariance matrix of X and Y. Here, we can estimate principal components of C_{YY} by analyzing $C_{YY.X}$ and $C_{YY.\bar{X}}$ separately and the factors extracted from $C_{YY.\bar{X}}$ are free from influences of external criteria [174].

9.4.2 Local PCA with External Criteria

We consider a fuzzy clustering algorithm that extracts local principal components uncorrelated to external criteria [120]. In the technique, **FCRM** is used not only for partitioning a data set into fuzzy clusters but also for removing influences of external criteria from fuzzy scatter matrices. Solving the eigenvalue problems of the modified fuzzy scatter matrices, local principal components uncorrelated to external criteria are derived.

In order to use the switching regression model as the preprocessing of PCA, we use the centered data model for **FCRM** presented in Section 8.1.3.

$$y_{ik}^j = \sum_{\ell=1}^{p} x_{ik}^\ell \beta_{ij}^\ell + e_{ik}^j, \tag{9.79}$$

where $x_{ik}^\ell = x_k^\ell - v_{xi}^\ell$ and $y_{ik}^j = y_k^j - v_{yi}^j$. $v_{xi} = (v_{xi}^1, \dots, v_{xi}^p)^\top$ and $v_{yi} = (v_{yi}^1, \dots, v_{yi}^r)^\top$ are the center (mean vector) of cluster i. Necessary condition for the optimality reduces the optimal B_i as

$$\mathsf{B}_i = (\mathsf{X}_i^\top \mathsf{U}_i^m \mathsf{X}_i)^{-1}\mathsf{X}_i^\top \mathsf{U}_i^m \mathsf{Y}_i. \tag{9.80}$$

Using the local linear models, the data matrix to be analyzed is decomposed in each cluster as

$$\begin{aligned} \mathsf{Y}_i &= \mathsf{X}_i\mathsf{B}_i + (\mathsf{Y}_i - \mathsf{X}_i\mathsf{B}_i) \\ &= \mathsf{Y}_{X_i} + \mathsf{Y}_{\bar{X}_i}. \end{aligned} \tag{9.81}$$

Then, the generalized fuzzy scatter matrix of cluster i is also decomposed as

$$\begin{aligned} S_{fi} &= \mathsf{Y}_i^\top \mathsf{U}_i^m \mathsf{Y}_i \\ &= \mathsf{Y}_i^\top \mathsf{U}_i^m (\mathsf{Y}_{X_i} + \mathsf{Y}_{\bar{X}_i}) \\ &= \mathsf{Y}_i^\top \mathsf{U}_i^m \mathsf{X}_i \mathsf{B}_i + \{\mathsf{Y}_i^\top \mathsf{U}_i^m \mathsf{Y}_i - \mathsf{Y}_i^\top \mathsf{U}_i^m \mathsf{X}_i \mathsf{B}_i\} \\ &= S_{fi.X} + S_{fi.\bar{X}}, \end{aligned} \tag{9.82}$$

where

$$S_{fi.X} = \mathsf{Y}_i^\top \mathsf{U}_i^m \mathsf{X}_i (\mathsf{X}_i^\top \mathsf{U}_i^m \mathsf{X}_i)^{-1} \mathsf{X}_i^\top \mathsf{U}_i^m \mathsf{Y}_i, \tag{9.83}$$

$$S_{fi.\bar{X}} = \mathsf{Y}_i^\top \mathsf{U}_i^m \mathsf{Y}_i - \mathsf{Y}_i^\top \mathsf{U}_i^m \mathsf{X}_i (\mathsf{X}_i^\top \mathsf{U}_i^m \mathsf{X}_i)^{-1} \mathsf{X}_i^\top \mathsf{U}_i^m \mathsf{Y}_i. \tag{9.84}$$

U_i is the $N \times N$ diagonal matrix whose k-th diagonal element is u_{ki}. Here, $S_{fi.X}$ is closely related to X_i while $S_{fi.\bar{X}}$ is independent of X_i. So, the local principal components uncorrelated to external criteria are derived by solving the eigenvalue problems of $S_{fi.\bar{X}}$.

9.5 Fuzzy Local Independent Component Analysis

Independent component analysis (ICA) [18, 66, 89] is an unsupervised technique, which in many cases characterizes data sets in a natural way, and have been applied to blind source separation (BSS) problems. ICA tries to represent data sets in term of statistically independent variables using higher order statistics as the criterion instead of the second order statistics used in PCA. The mutual dependence of components is classically measured by their non-gaussianity. Maximizing non-gaussianity gives us one of independent components and the procedures are closely related to projection pursuit [65] that is also a technique developed in statistics for finding "interesting" features of multivariate data. In projection pursuit, the goal is to find the one-dimensional projections of multivariate data, which have "interesting" distributions for visualization purposes. Typically, the interestingness is measured by the non-gaussianity. Therefore, the basis vectors of ICA should be especially useful in projection pursuit and in extracting characteristic features from natural data [79].

In this section, we introduce a local ICA technique based on linear fuzzy clustering after a brief review of ICA formulation.

9.5.1 ICA Formulation and Fast ICA Algorithm

Denote that x is a p dimensional observed data vector and s is a q dimensional source signal vector corresponding to the observed data with $q \leq p$,

$$x = (x^1, x^2, \ldots, x^p)^\top,$$
$$s = (s^1, s^2, \ldots, s^q)^\top.$$

When the elements of source signals $(s^1, s^2, \ldots, s^q)$ are mutually statistically independent and have zero-means, observed data are assumed to be linear mixtures of s^i as follows:

$$x = As, \tag{9.85}$$

where unknown $p \times q$ matrix A is called a mixing matrix. The goal of ICA is to estimate source signals s^i, $i = 1, \ldots, q$ and mixing matrix A using only observed data x.

A lot of algorithms for ICA have been proposed, including neural learning algorithms [77]. It is useful to apply a preprocessing of whitening or sphering by using PCA before applying the ICA algorithms [15, 18, 122]. In the preprocessing, observed data x are transformed into linear combinations z,

$$z = P^\top x,$$

such that their elements z^i, $i = 1, \ldots, q$ are mutually uncorrelated and all have unit variance. This preprocessing implies that correlation matrix $E\{zz^\top\}$ is equal to unit matrix I_q, and is usually performed by PCA.

After transformation, we have

$$z = P^\top x = P^\top As = Ws,$$

where $W = P^\top A$ is an orthogonal matrix due to the assumption. Thus we can reduce the problem of finding arbitrary full-rank matrix A to a simpler problem of finding an orthogonal matrix W, which gives $s = W^\top z$.

Fast ICA proposed by Hyvärinen and Oja [67] is a useful algorithm that is very simple, does not depend on any user-defined parameters, and is fast to converge to the most accurate solution. In the original fast ICA formulation, Hyvärinen and Oja used non-gaussianity as the measure of the mutual dependence of reconstructed variables, and proposed the following objective function to be minimized or maximized:

$$J_{\text{fica}}(w) = E\{(w^\top z)^4\} - 3\|w\|^4 + F(\|w\|^2), \tag{9.86}$$

where q dimensional vector w is a colmun of mixing matrix W. $E\{\cdot\}$ denotes sample mean. $E\{(w^\top z)^4\} - 3\|w\|^4$ is the fourth-order cumulant or kurtosis that measures the gaussianity of distribution. Maximizing the non-gaussianity of reconstructed signals gives us one of independent components. The third term denotes the constraint of w such that $\|w\|^2 = 1$.

The fast ICA Algorithm that uses fixed-point iteration [67] is represented as follows:

Algorithm FICA: *A Fast Fixed-Point Algorithm for ICA*

FICA1. Take a random initial weight vector $w(0)$ of norm 1. Let $r = 1$.
FICA2. Update $w(r)$ using (9.87).

$$w(r) = E\{z(w(r-1)^\top z)^3\} - 3w(r-1) \tag{9.87}$$

FICA3. Divide $w(r)$ by its norm.
FICA4. If $|w(r)^\top w(r-1)|$ is enough close to 1, stop: otherwise Let $r = r+1$ and return to FICA2.
End FICA.

Vectors $w(r)$ obtained by the algorithm constitute the columns of orthogonal mixing matrix W. To estimate q independent components, we need to run this algorithm q times. We can estimate the independent components one by one by adding projection operation in the beginning of FICA3.

9.5.2 Fuzzy Local ICA with FCV Clustering

In spite of their usefulness, linear ICA models are often too simple to describe real world data. Karhunen *et al.* proposed local ICA models [78] that were used in connection with suitable clustering algorithms. In these local ICA models, data are grouped in several clusters based on similarities between observed data ahead of preprocessing of linear ICA using some clustering algorithms such as K-means. Observed data, however, are assumed to be linear combinations of source signals in BSS problems. So, the clustering methods that partition data into some spherical clusters, like by K-means, are not suitable for extraction of local independent components. Taking similarities of source signals into account, we should partition data into several linear clusters.

Fuzzy fast ICA [59] is a fuzzy version of the fast ICA algorithm that can handle fuzziness in the iterative algorithm by using the FCV clustering as preprocessing. The FCV clustering simultaneously performs fuzzy clustering and PCA. The principal subspace of each cluster is estimated as the prototypical linear variety of dimension q that passes through a point (cluster center) v_i and is spanned by linearly independent vectors $a_{i1}, \ldots, a_{iq}$.

Considering memberships $u_{ki}, k = 1, \ldots, N; i = 1, \ldots, c$, whitening of data is performed in each cluster as follows:

Step 1. Project data onto the prototypical linear variety in cluster i,

$$\tilde{x}_{ik} = \mathsf{A}_i^\top (x_k - v_i),$$

where

$$\mathsf{A}_i = (a_{i1}, \ldots, a_{iq}). \tag{9.88}$$

Step 2. Transform the projected data to have unit variance,

$$f_{ik} = D_{\sigma_i}^{-1} \tilde{x}_{ik},$$

where

$$D_{\sigma_i} = \mathrm{diag}\left(\sigma_i^1, \ldots, \sigma_i^q\right),$$

$$\sigma_i^\ell = \sqrt{E\{(\tilde{x}_{i\cdot}^\ell)^2\}} = \sqrt{\frac{\sum_{k=1}^{N} (u_{ki})^m (\tilde{x}_{ik}^\ell)^2}{\sum_{k=1}^{N} (u_{ki})^m}}.$$

By using this preprocessing, observed data x_k are transformed to f_{ik} that satisfy $E\{f_{i\cdot} f_{i\cdot}^\top\} = I_q$ in cluster i. Here, we consider the generalized weight of $(u_{ki})^m$ that is used in FCM-based local PCA. In the case of the entropy-based method, we put $m = 1$.

For measuring the non-gaussianity of linear combination $w_i^\top f_{i\cdot}$ in cluster i, we use fuzzy kurtosis and denote the objective function for fuzzy ICA as follows [57, 59]:

$$\begin{aligned} J_{\text{ffica}} &= E\{(w_i^\top f_{i\cdot})^4\} - 3\|w_i\|^4 + F(\|w_i\|^2) \\ &= \frac{\sum_{k=1}^{N} (u_{ki})^m (w_i^\top f_{ik})^4}{\sum_{k=1}^{N} (u_{ki})^m} - 3\|w_i\|^4 + F(\|w_i\|^2), \end{aligned} \tag{9.89}$$

$E\{(w_i^\top f_{i\cdot})^4\} - 3\|w_i\|^4$ is the fuzzy kurtosis in which the average is calculated by considering generalized membership weights. When the fuzzy kurtosis is zero, $w_i^\top f_{i\cdot}$ has gaussian distribution in fuzzy cluster i. So, minimization or maximization of this objective function derives $w_i^\top f_{i\cdot}$ that has non-gaussian distribution. The third term denotes the constraint of w_i such that $\|w_i\|^2 = 1$.

In this way, this method applies ICA to fuzzily partitioned data taking account of the memberships obtained in preprocessing by **FCV**.

The fuzzy fast ICA Algorithm that uses a fixed-point algorithm is as follows:

Algorithm FFICA: *A Fast Fixed-Point Algorithm for Fuzzy ICA*

FFICA1. Take a random initial weight vector $w_i(0)$ of norm 1. Let $r = 1$.
FFICA2. Update $w_i(r)$ using (9.90).

$$w_i(r) = E\{f_{i\cdot}(w_i(r-1)^\top f_{i\cdot})^3\} - 3w_i(r-1), \tag{9.90}$$

where

$$E\{f_{i\cdot}(w_i(r-1)^\top f_{i\cdot})^3\} = \frac{\sum_{k=1}^{N}(u_{ki})^m f_{ik}(w_i(r-1)^\top f_{ik})^3}{\sum_{k=1}^{N}(u_{ki})^m}. \quad (9.91)$$

FFICA3. Divide $w_i(r)$ by its norm.
FFICA4. If $|w_i(r)^\top w_i(r-1)|$ is enough close to 1, stop: otherwise Let $r = r+1$ and return to FFICA2.
End FFICA.

The vectors $w_i(r)$ obtained by the algorithm constitute the columns of the orthogonal mixing matrix W_i in cluster i. To estimate q independent components, we need to run this algorithm q times in each cluster. We can estimate the independent components one by one by adding the projection operation in the beginning of FFICA3 as follows:

$$w_i(r) = w_i(r) - \bar{W}_i\bar{W}_i^\top w_i(r),$$

where $\bar{W}_i$ is a matrix whose columns are the previously found columns of W_i.

9.5.3 An Illustrative Example

Let us demonstrate the performance of the **FFICA** algorithm using a blind source separation (BSS) problem. Assume that three microphones observed mixed speech signals that were the linear mixtures of the voices of two persons speaking simultaneously. The goal of BSS is to decompose the mixed signals into the two voice sources. It is usually assumed that all of the data were mixed by using a unique mixing matrix. This assumption, however, is not always satisfied in real world data. We consider a case where the mixing matrix was changed in the middle of the observation process because the speakers moved from their original places to others. The mixed signals observed by three microphones were made by using two mixing matrices Q_1 and Q_2,

$$Q_1 = \begin{pmatrix} 0.5 & 0.5 \\ 0.7 & 0.3 \\ 0.2 & 0.8 \end{pmatrix}, \quad Q_2 = \begin{pmatrix} 0.5 & 0.5 \\ 0.6 & 0.4 \\ 0.8 & 0.2 \end{pmatrix}.$$

The first half of the data was mixed by Q_1 while the second half was mixed by Q_2. Figures 9.10 and 9.11 show the 3-D plots of the observed data and the observed mixed signals, respectively. In Fig. 9.10, the data form the two planes. The experiment was carried out to reconstruct the two original speech signals using the fuzzy fast ICA algorithm that can consider the two local linear structures given by two different mixing matrices.

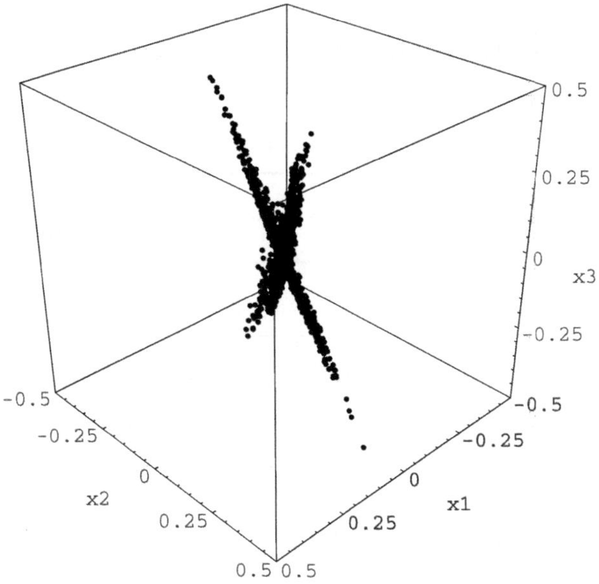

Fig. 9.10. 3D plots of mixed speech signals

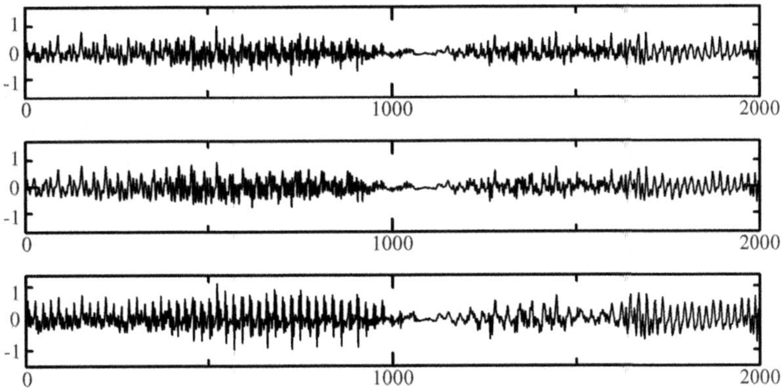

Fig. 9.11. Three mixed speech signals

In the preprocessing stage, we applied the standard **FCV** algorithm with $c = 2$ and $m = 2$. The data was partitioned into two clusters, one consists of the first half of the data and another consists of the second half. After the preprocessing stage, we performed the fuzzy fast ICA algorithm in each cluster using the memberships derived by **FCV**. Figure 9.12 shows the comparisons between the original and the reconstructed speech signals of the two persons, where the reconstructed signal series of the second cluster were connected to the tails of that of the first cluster. Both original speech signals were reconstructed

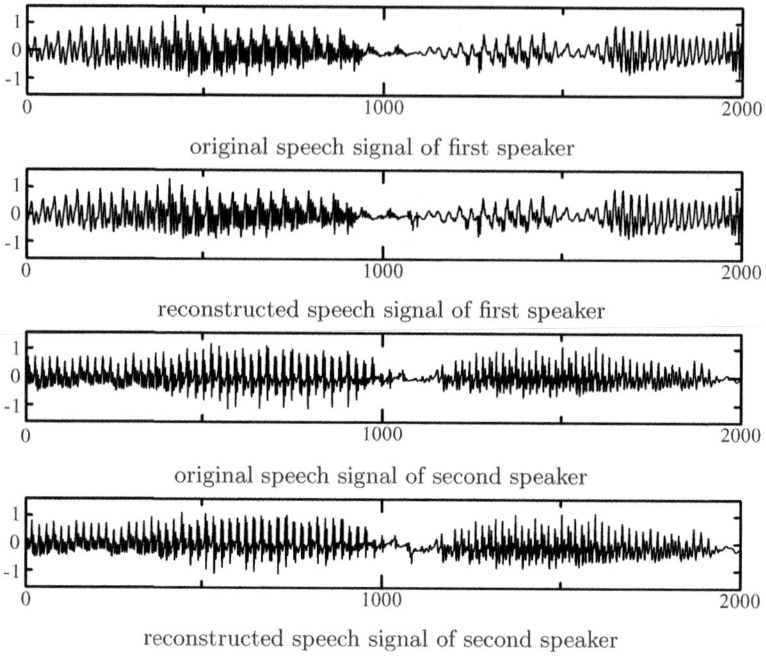

Fig. 9.12. Comparison between original and reconstructed speech signals

successfully. It is because **FCV** separated data into two parts in response to the change of the mixing matrix. So, it can be said that the local ICA method partitions observed data based on similarities of source signals.

9.6 Fuzzy Local ICA with External Criteria

From the view point of knowledge discovery in databases, it is useful to extract independent components uncorrelated to some external criteria. In this section, we introduce an enhanced technique for local independent component analysis [57], in which preprocessing is performed by using the fuzzy clustering algorithm that extracts local principal components uncorrelated to external criteria presented in Section 9.4.

9.6.1 Extraction of Independent Components Uncorrelated to External Criteria

We consider to apply the ICA algorithm to principal components extracted by Yanai's technique (see Section 9.4.1). Let $\mathsf{P} = (p_1, p_2, \ldots, p_q)$ be the $q \times q$ matrix composed of q principal eigenvectors of $C_{YY.\bar{X}}$. Normalized data matrix $\mathsf{Z}_{\bar{X}}$ to be analyzed is calculated as

$$\begin{aligned} \mathsf{Z}_{\bar{X}} &= \mathsf{Y}_{\bar{X}}\mathsf{P} \\ &= \left(I_N - \mathsf{X}(\mathsf{X}^\top\mathsf{X})^{-1}\mathsf{X}^\top\right)\mathsf{YP}. \end{aligned} \tag{9.92}$$

Here,

$$C_{XZ_{\bar{X}}} = \frac{1}{N}\mathsf{X}^\top\left(I_N - \mathsf{X}(\mathsf{X}^\top\mathsf{X})^{-1}\mathsf{X}^\top\right)\mathsf{YP} = O$$

implies that the covariance matrix of $\mathsf{Z}_{\bar{X}}$ and X is O (zero matrix), i.e., the normalized data are uncorrelated to external criteria.

If we apply the ICA algorithm to this normalized data matrix by multiplying orthogonal matrix W, reconstructed data matrix S are derived as

$$\mathsf{S} = \mathsf{Z}_{\bar{X}}W, \tag{9.93}$$

and the covariance matrix of S and X is calculated as

$$C_{XS} = C_{XZ_{\bar{X}}}W = O. \tag{9.94}$$

Therefore, the independent components are also uncorrelated to external criteria. In this way, we can extract independent components uncorrelated to some external criteria by removing influences of external criteria in preprocessing by Yanai's technique.

9.6.2 Extraction of Local Independent Components Uncorrelated to External Criteria

Next, we estimate local independent components uncorrelated to some external criteria by applying the fuzzy ICA algorithm to local principal components extracted in Section 9.4.2. Using the local linear model, the data matrix to be analyzed is decomposed in cluster i as

$$\begin{aligned} \mathsf{Y}_i &= \mathsf{X}_i\mathsf{B}_i + (\mathsf{Y}_i - \mathsf{X}_i\mathsf{B}_i) \\ &= \mathsf{Y}_{X_i} + \mathsf{Y}_{\bar{X}_i}. \end{aligned} \tag{9.95}$$

Then, the fuzzy scatter matrix of cluster i is also decomposed as

$$\begin{aligned} S_{fi} &= \mathsf{Y}_i^\top\mathsf{U}_i^m\mathsf{Y}_i \\ &= \mathsf{Y}_i^\top\mathsf{U}_i^m(\mathsf{Y}_{X_i} + \mathsf{Y}_{\bar{X}_i}) \\ &= \mathsf{Y}_i^\top\mathsf{U}_i^m\mathsf{X}_i\mathsf{B}_i + \{\mathsf{Y}_i^\top\mathsf{U}_i^m\mathsf{Y}_i - \mathsf{Y}_i^\top\mathsf{U}_i^m\mathsf{X}_i\mathsf{B}_i\} \\ &= S_{fi.X} + S_{fi.\bar{X}}, \end{aligned} \tag{9.96}$$

where

$$S_{fi.X} = \mathsf{Y}_i^\top\mathsf{U}_i^m\mathsf{X}_i(\mathsf{X}_i^\top\mathsf{U}_i^m\mathsf{X}_i)^{-1}\mathsf{X}_i^\top\mathsf{U}_i^m\mathsf{Y}_i, \tag{9.97}$$

$$S_{fi.\bar{X}} = \mathsf{Y}_i^\top\mathsf{U}_i^m\mathsf{Y}_i - \mathsf{Y}_i^\top\mathsf{U}_i^m\mathsf{X}_i(\mathsf{X}_i^\top\mathsf{U}_i^m\mathsf{X}_i)^{-1}\mathsf{X}_i^\top\mathsf{U}_i^m\mathsf{Y}_i. \tag{9.98}$$

U_i is the $N \times N$ diagonal matrix whose k-th diagonal element is u_{ki}. Here, $S_{fi.X}$ is closely related to X_i while $S_{fi.\bar{X}}$ is independent of X_i. So, the local principal components uncorrelated to the external criteria are derived by solving the eigenvalue problems of $S_{fi.\bar{X}}$, and local independent components are estimated from the local principal components by using the fuzzy fast ICA algorithm [57].

9.7 Fuzzy Clustering-Based Variable Selection in Local PCA

In the analysis of large scale data sets, it is often the case that they include unnecessary variables, which is not informative for modeling. There are two approaches for estimating responsibility of variables in PCA. In dimension reduction tasks, the goal is to select items or variables so as to keep the original information as well as possible. When several variables are mutually dependent, we can estimate the data substructure even if we eliminate some of the redundant variables. In order to select the redundant variables, several criteria for variable selection in PCA have been proposed [75, 153, 183]. When we want to obtain the best subset of variables, we should search for the subset which has the largest (or smallest) criterion value among all possible subsets. On the other hand, in data mining tasks, the goal is to extract association rules and we wish to select the variables that are mutually dependent, i.e., we should eliminate variables that have no responsibility for estimation of principal subspace.

In this section, we consider an approach for variable selection in linear fuzzy clustering, which selects the variables that have close relationship with local principal components, by introducing the mechanism of variable selection into the iterative algorithm of linear fuzzy clustering [58].

9.7.1 Linear Fuzzy Clustering with Variable Selection

In order to eliminate unnecessary variables, memberships of variables are introduced into the least squares criterion for local PCA. Using two types of memberships, the objective function with the standard fuzzification approach is defined as follows:

$$J_{\text{fcvvs}}(U, W, V, F, A) = \sum_{i=1}^{c}\sum_{k=1}^{N}(u_{ki})^m \sum_{j=1}^{p}(w_{ji})^t \left(x_k^j - \sum_{\ell=1}^{q} f_{ik}^{\ell} a_{i\ell}^j - v_i^j \right)^2 , \tag{9.99}$$

where w_{ji} represents the degree of membership of variable j to cluster i and W is the total set of w_{ji}. The weighting exponent $t (t > 1)$ plays a role for fuzzification of membership degrees of variables. If variable j has no useful information for estimating the i-th prototypical linear variety, w_{ji} has small value and variable j is ignored in calculation of clustering criteria in cluster i.

When we use the entropy-based method, the objective function is given as

$$J_{\text{efcvvs}}(U,W,V,F,A) = \sum_{i=1}^{c}\sum_{k=1}^{N} u_{ki} \sum_{j=1}^{p} w_{ji} \left(x_k^j - \sum_{\ell=1}^{q} f_{ik}^{\ell} a_{i\ell}^j - v_i^j \right)^2$$
$$+\nu \sum_{i=1}^{c}\sum_{k=1}^{N} u_{ki} \log u_{ki}$$
$$+\eta \sum_{i=1}^{c}\sum_{j=1}^{p} w_{ji} \log w_{ji}. \quad (9.100)$$

The second entropy term plays a role for fuzzification of memberships of variables, and the degree of fuzziness is controlled by the fuzzifier η.

To obtain a unique solution, the objective function is minimized under the constraints (9.6)-(9.8) and $\mathsf{A}_i^\top \mathsf{A}_i$ is diagonal. For the entropy-based method, we put $m = 1$. Additionally, the constraint for memberships of variables is given as

$$\sum_{j=1}^{p} w_{ji} = 1 \quad ; i = 1, \dots, c, \quad (9.101)$$

and the additional memberships represent relative responsibilities of variables.

We can use either $J(U,W,V,F,A) = J_{\text{fcvvs}}(U,W,V,F,A)$ or $J(U,W,V,F,A) = J_{\text{efcvvs}}(U,W,V,F,A)$, and the algorithm for FCV with variable selection can be written as follows:

Algorithm FCV-VS: *Fuzzy c-Varieties with Variable Selection* [58]

FCV-VS1. [Generate initial value:] Generate initial values for $\bar{U} = (\bar{\mathsf{U}}_1, \dots, \bar{\mathsf{U}}_c)$, $\bar{W} = (\bar{\mathsf{W}}_1, \dots, \bar{\mathsf{W}}_c)$, $\bar{V} = (\bar{v}_1, \dots, \bar{v}_c)$, $\bar{F} = (\bar{\mathsf{F}}_1, \dots, \bar{\mathsf{F}}_c)$ and $\bar{A} = (\bar{\mathsf{A}}_1, \dots, \bar{\mathsf{A}}_c)$, and normalize them so that they satisfy the constraints (9.6)-(9.8), (9.101) and $\bar{\mathsf{A}}_i^\top \bar{\mathsf{A}}_i$ is diagonal.

FCV-VS2. [Find optimal U:] Calculate

$$\bar{U} = \arg\min_{U \in \mathcal{U}_f} J(U, \bar{W}, \bar{V}, \bar{F}, \bar{A}). \quad (9.102)$$

FCV-VS3. [Find optimal W:] Calculate

$$\bar{W} = \arg\min_{W \in \mathcal{W}_f} J(\bar{U}, W, \bar{V}, \bar{F}, \bar{A}), \quad (9.103)$$

where

$$\mathcal{W}_f = \{ W = (w_{ji}) : \sum_{k=1}^{p} w_{ki} = 1,\ 1 \le i \le c;$$
$$w_{ji} \in [0,1],\ 1 \le j \le p,\ 1 \le i \le c \}. \quad (9.104)$$

FCV-VS4. [Find optimal V:] Calculate

$$\bar{V} = \arg\min_{V} J(\bar{U}, \bar{W}, V, \bar{F}, \bar{A}). \quad (9.105)$$

FCV-VS5. [Find optimal F:] Calculate

$$\bar{F} = \arg\min_{F} J(\bar{U}, \bar{W}, \bar{V}, F, \bar{A}), \tag{9.106}$$

and normalize them so that they satisfy the constraints (9.6) and (9.7).
FCV-VS6. [Find optimal A:] Calculate

$$\bar{A} = \arg\min_{A} J(\bar{U}, \bar{W}, \bar{V}, \bar{F}, A), \tag{9.107}$$

and transform them so that each $\mathsf{A}_i^\top \mathsf{A}_i$ is diagonal.
FCV-VS7. [Test convergence:] If all parameters are convergent, stop; else go to FCV-VS2.
End FCV-VS.

The orthogonal matrices in FCV-VS1, FCV-VS5 and FCV-VS6 are obtained by such a technique as Gram-Schmidt's orthogonalization.

To derive the optimal values of parameters, the objective function (9.99) is rewritten as follows:

$$J_{\text{fcvvs}} = \sum_{i=1}^{c}\sum_{j=1}^{p}(w_{ji})^t(\tilde{x}_j - \mathsf{F}_i\tilde{a}_i^j - 1_N v_i^j)^\top \mathsf{U}_i^m(\tilde{x}_j - \mathsf{F}_i\tilde{a}_i^j - 1_N v_i^j), \tag{9.108}$$

where

$$\mathsf{X} = (\tilde{x}_1, \ldots, \tilde{x}_p),$$
$$\mathsf{A}_i = (\tilde{a}_i^1, \ldots, \tilde{a}_i^p)^\top.$$

From $\partial J_{\text{fcvvs}}/\partial \tilde{a}_i^j = 0_q$ and $\partial J_{\text{fcvvs}}/\partial v_i^j = 0$, we have

$$\tilde{a}_i^j = (\mathsf{F}_i^\top \mathsf{U}_i^m \mathsf{F}_i)^{-1}\mathsf{F}_i^\top \mathsf{U}_i^m(\tilde{x}_j - 1_N v_i^j), \tag{9.109}$$
$$v_i^j = (1_N^\top \mathsf{U}_i^m 1_N)^{-1} 1_N^\top \mathsf{U}_i^m(\tilde{x}_j - \mathsf{F}_i\tilde{a}_i^j). \tag{9.110}$$

In the same way, (9.99) is equivalent to

$$J_{\text{fcvvs}} = \sum_{i=1}^{c}\sum_{k=1}^{N}(u_{ki})^m(x_k - \mathsf{A}_i f_{ik} - v_i)^\top \mathsf{W}_i^t(x_k - \mathsf{A}_i f_{ik} - v_i), \tag{9.111}$$

and $\partial J_{\text{fcvvs}}/\partial f_{ik} = 0_q$ yields

$$f_{ik} = (\mathsf{A}_i^\top \mathsf{W}_i^t \mathsf{A}_i)^{-1}\mathsf{A}_i^\top \mathsf{W}_i^t(x_k - v_i), \tag{9.112}$$

where $\mathsf{W}_i = \text{diag}(w_{1i}, \ldots, w_{pi})$.

Note that we can derive the updating rules for the entropy-based method by setting $m = 1$ and $t = 1$.

The membership values u_{ki} are given as

$$u_{ki} = \left[\sum_{j=1}^{c} \left(\frac{D(x_k, P_i)}{D(x_k, P_j)} \right)^{\frac{1}{m-1}} \right]^{-1}, \tag{9.113}$$

for $J_{\text{fcvvs}}(U, W, V, F, A)$ and

$$u_{ki} = \frac{\exp\left(-\dfrac{D(x_k, P_i)}{\nu} \right)}{\displaystyle\sum_{j=1}^{c} \exp\left(-\dfrac{D(x_k, P_j)}{\nu} \right)}, \tag{9.114}$$

for $J_{\text{efcvvs}}(U, W, V, F, A)$ where

$$D(x_k, P_i) = \sum_{j=1}^{p} (w_{ji})^t \left(x_k^j - \sum_{\ell=1}^{q} f_{ik}^{\ell} a_{i\ell}^j - v_i^j \right)^2. \tag{9.115}$$

On the other hand, the membership values w_{ji} are given as

$$w_{ji} = \left[\sum_{\ell=1}^{p} \left(\frac{E(\tilde{x}_j, P_i)}{E(\tilde{x}_\ell, P_i)} \right)^{\frac{1}{t-1}} \right]^{-1}, \tag{9.116}$$

for $J_{\text{fcvvs}}(U, W, V, F, A)$ and

$$w_{ji} = \frac{\exp\left(-\dfrac{E(\tilde{x}_j, P_i)}{\eta} \right)}{\displaystyle\sum_{\ell=1}^{p} \exp\left(-\dfrac{E(\tilde{x}_\ell, P_i)}{\eta} \right)}, \tag{9.117}$$

for $J_{\text{efcvvs}}(U, W, V, F, A)$ where

$$E(\tilde{x}_j, P_i) = \sum_{k=1}^{N} (u_{ki})^m \left(x_k^j - \sum_{\ell=1}^{q} f_{ik}^{\ell} a_{i\ell}^j - v_i^j \right)^2. \tag{9.118}$$

In this way, local subspaces are estimated ignoring unnecessary variables that have small memberships.

9.7.2 Graded Possibilistic Variable Selection

When the number of variables p is large, the values will be very small because of the constraint of (9.101). It is often difficult to interpret the absolute responsibility of a variable from its responsibility value. The deficiency comes from the fact that it imposes the same constraint with the conventional memberships on variable selection parameters. In the mixed c-means clustering, Pal *et al.* [127]

proposed to relax the constraint (row sum = 1) on the typicality values but retain the column constraint on the membership values.

In the possibilistic approach [87], memberships can be regarded as the probability that an experimental outcome coincides with one of mutually independent events. However, it is possible that sets of events are neither mutually independent nor completely mutually exclusive. Then, Masulli and Rovetta [97] proposed the graded possibilistic approach to clustering, in which soft transition of memberships from probabilistic to possibilistic constraint is performed by using the graded possibilistic constraint. In [58], absolute typicalities of variables are estimated by using the graded possibilistic clustering.

9.7.3 An Illustrative Example

A numerical experiment was performed using an artificial data set [58]. Table 9.2 shows the coordinates of the samples. Samples 1-12 form the first group, in which x^1, x^2 and x^3 are linearly related, i.e., samples are distributed forming a line in the 3-D space. However, x^4 and x^5 are random variables. So, x^1, x^2 and x^3 should be selected in the group and we can capture the local linear structure by eliminating x^4 and x^5. On the other hand, samples 13-24 form the second group, in which x^2, x^3 and x^4 are linearly related, but x^1 and x^5 are

Table 9.2. Artificial data set including unnecessary variables

sample	x^1	x^2	x^3	x^4	x^5
1	0.000	0.000	0.250	**0.143**	**0.365**
2	0.091	0.091	0.295	**0.560**	**0.605**
3	0.182	0.182	0.341	**0.637**	**0.001**
4	0.273	0.273	0.386	**0.529**	**0.557**
5	0.364	0.364	0.432	**0.949**	**0.195**
6	0.455	0.455	0.477	**0.645**	**0.206**
7	0.545	0.545	0.523	**0.598**	**0.026**
8	0.636	0.636	0.568	**0.616**	**0.729**
9	0.727	0.727	0.614	**0.004**	**0.407**
10	0.818	0.818	0.659	**0.255**	**0.641**
11	0.909	0.909	0.705	**0.088**	**0.244**
12	1.000	1.000	0.750	**0.589**	**0.213**
13	**0.199**	0.750	0.250	0.000	**0.321**
14	**0.411**	0.705	0.295	0.091	**0.167**
15	**0.365**	0.659	0.341	0.182	**0.419**
16	**0.950**	0.614	0.386	0.273	**0.109**
17	**0.581**	0.568	0.432	0.364	**0.561**
18	**0.323**	0.523	0.477	0.455	**0.127**
19	**0.899**	0.477	0.523	0.545	**0.349**
20	**0.399**	0.432	0.568	0.636	**0.100**
21	**0.249**	0.386	0.614	0.727	**0.682**
22	**0.214**	0.341	0.659	0.818	**0.714**
23	**0.838**	0.295	0.705	0.909	**0.605**
24	**0.166**	0.250	0.750	1.000	**0.244**

Table 9.3. Memberships of variables and local principal component vectors with standard fuzzification

variable	w_{ji}		a_i	
	$i = 1$	$i = 2$	$i = 1$	$i = 2$
x^1	**0.318**	0.018	**1.083**	0.036
x^2	**0.318**	**0.316**	**1.083**	**0.539**
x^3	**0.318**	**0.316**	**0.541**	**-0.539**
x^4	0.020	**0.316**	-0.236	**-1.079**
x^5	0.026	0.035	0.007	-0.291

random variables, i.e., the local structures must be captured by classifying not only samples but also variables.

Applying the **FCV-VS** algorithm with the standard fuzzification method, the samples were partitioned into two clusters using the probabilistic constraint. The model parameters were set as $m = 2.0$, $t = 2.0$, $p = 1$. In the sense of maximum membership, the first cluster included samples 1-12, while the second cluster included the remaining samples. Table 9.3 shows the derived memberships of variables and local principal component vectors. x^4 and x^5 were eliminated in the first cluster and a_1 revealed the relationship among x^1, x^2 and x^3. On the other hand, in the second cluster, small memberships were assigned to x^1 and x^5, and a_2 represented the local structure of the second group.

The clustering result indicates that the proposed membership w_{ji} is useful for evaluating the typicality of variables in local linear model estimation where x^1 and x^4 are significant only in the first and second cluster, respectively. Additionally, the typicality values also play a role for rejecting the influences of noise variable (x^5) because x^5 belongs to neither of two clusters. In this way, the row sum constraints (9.101) give the memberships a different role from the conventional column constraints that forces each samples to belong at least one cluster.

References

1. Anderberg, M.R.: Cluster Analysis for Applications. Academic Press, New York (1973)
2. Ball, G.H., Hall, D.J.: A clustering technique for summarizing multivariate data. Behavioral Science 12, 153–155 (1967)
3. Ben-Hur, A., Horn, D., Siegelmann, H.T., Vapnik, V.: A support vector clustering method. In: Proc. of ICPR 2000, pp. 724–727 (2000)
4. Ben-Hur, A., Horn, D., Siegelmann, H.T., Vapnik, V.: Support vector clustering. Journal of Machine Learning Research 2, 125–137 (2001)
5. Bezdek, J.C.: Fuzzy Mathematics in Pattern Classification, Ph.D. Thesis, Cornell Univ., Ithaca, NY (1973)
6. Bezdek, J.C.: Pattern Recognition with Fuzzy Objective Function Algorithms. Plenum Press (1981)
7. Bezdek, J.C., Coray, C., Gunderson, R., Watson, J.: Detection and characterization of cluster substructure 2. fuzzy c-varieties and convex combinations thereof. SIAM J. Appl. Math. 40(2), 358–372 (1981)
8. Bezdek, J.C., Keller, J., Krishnapuram, R., Pal, N.R.: Fuzzy Models and Algorithms for Pattern Recognition and Image Processing. Kluwer, Boston (1999)
9. Bezdek, J.C., Kuncheva, L.I.: Nearest prototype classifier designs: An experimental study. Int. J. of Intelligent Systems 16, 1445–1473 (2001)
10. Bishop, C.M.: Neural Networks for Pattern Recognition. Clarendon Press (1995)
11. Bloomfield, P., Steiger, W.L.: Least Absolute Deviations: Theory, Applications, and Algorithms. Birkhäuser Verlag, Boston (1983)
12. Bobrowski, L., Bezdek, J.C.: c-Means Clustering with the l_1 and l_∞ Norms. IEEE Trans. on Syst., Man, and Cybern. 21(3), 545–554 (1991)
13. Bond, J., Michailidis, G.: Homogeneity analysis in Lisp-Stat. Journal of Statistical Software 1(2) (1996)
14. Burges, C.J.C.: A tutorial on support vector machines for pattern recognition. Data Mining and Knowledge Discovery 2, 121–167 (1998)
15. Cardoso, J.-F.: Eigen-structure of the fourth-order cumulant tensor with application to the blind source separation problem. In: Proc. IEEE Int. Conf. on Acoustics, Speech, and Signal Processing, pp. 2655–2658 (1990)
16. Celeux, G., Govaert, G.: A classification EM algorithm for clustering and two stochastic versions. Computational Statistics & Data Analysis 14, 315–332 (1992)
17. Celeux, G., Govaert, G.: Fuzzy clustering and mixture models. Compstat 94, 154–159 (1994)

18. Comon, P.: Independent component analysis - A new concept? Signal Processing 36, 287–314 (1994)
19. Cortes, C., Vapnik, V.: Support-vector network. Machine Learning 20, 273–297 (1995)
20. Cristianini, N., Shawe-Taylor, J.: An Introduction to Support Vector Machines and Other Kernel-Based Learning Methods. Cambridge University Press, Cambridge (2000)
21. Davé, R.N.: An adaptive fuzzy c-elliptotype clustering algorithm. Proc. of the North American Fuzzy Information Processing Society: Quarter Century of Fuzziness 1, 9–12 (1990)
22. Davé, R.N.: Characterization and detection of noise in clustering. Pattern Recognition Letters 12(11), 657–664 (1991)
23. Davé, R.N., Krishnapuram, R.: Robust clustering methods: A unified view. IEEE Trans. on Fuzzy Systems 5, 270–293 (1997)
24. de la Torre, F., Black, M.J.: A framework for robust subspace learning. International Journal of Computer Vision 54, 117–142 (2003)
25. Dempster, A.P., Laird, N.M., Rubin, D.B.: Maximum likelihood from incomplete data via the EM algorithm. J. R. Stat. Soc. B39, 1–38 (1977)
26. Dixon, J.K.: Pattern recognition with partly missing data. IEEE Trans. Syst., Man, Cybern. SMC-9, 617–621 (1979)
27. Draper, N.R., Smith, H.: Applied Regression Analysis. John Wiley & Sons, Inc., Chichester (1966)
28. Dumitrescu, D., Lazzerini, B., Jain, L.C.: Fuzzy Sets and Their Application to Clustering and Training. CRC Press, Baca Raton Florida (2000)
29. Duda, R.O., Hart, P.E.: Pattern Classification and Scene Analysis. John Wiley & Sons, Chichester (1973)
30. Duda, R.O., Hart, P.E., Stork, D.G.: Pattern Classification, 2nd edn. Wiley, New York (2001)
31. Dunn, J.C.: A fuzzy relative of the ISODATA process and its use in detecting compact well-separated clusters. J. of Cybernetics 3, 32–57 (1974)
32. Dunn, J.C.: Well-separated clusters and optimal fuzzy partitions. J. of Cybernetics 4, 95–104 (1974)
33. Elomaa, T., Rousu, J.: General and efficient multisplitting of numerical attributes. Machine Learning 36(3), 201–244 (1999)
34. Endo, Y., Haruyama, H., Okubo, T.: On some hierarchical clustering algorithms using kernel functions. In: Proc. of FUZZ-IEEE 2004, Budapest, Hungary, July 25-29, pp. 1–6 (2004) (CD-ROM)
35. Everitt, B.S.: Cluster Analysis, 3rd edn. Arnold, London (1993)
36. Fukunaga, K.: Statistical Pattern Recognition, 2nd edn. Academic Press, San Diego (1990)
37. Fukunaga, K., Olsen, D.R.: An algorithm for finding intrinsic dimensionality of data. IEEE Trans. on Computers C-20, 176–183 (1971)
38. Gath, I., Geva, A.B.: Unsupervised optimal fuzzy clustering. IEEE Trans. on Pattern Analysis and Machine Intelligence 11(7), 773–781 (1989)
39. Gath, I., Geva, A.B.: Fuzzy clustering for the estimation of the parameters of the components of mixtures of normal distributions. Pattern Recognition Letters 9(2), 77–86 (1989)
40. Geman, S., McClure, D.E.: Statistical methods for tomographic image reconstruction. Bulletin of International Statistical Institute LII-4, 5–21 (1987)
41. Gifi, A.: Nonlinear Multivariate Analysis. Wiley, Chichester (1990)

42. Girolami, M.: Mercer kernel based clustering in feature space. IEEE Trans. on Neural Networks 13(3), 780–784 (2002)
43. Golub, G.H., van Loan, C.F.: Matrix Computations, 2nd edn. The Johns Hopkins Univ. Press, Baltimore (1989)
44. Gu, Y., Hall, L.O.: Kernel based fuzzy ant clustering with partition validity. In: Proc. of 2006 IEEE International Conference on Fuzzy Systems, Sheraton Vancouver Wall Centre Hotel, Vancouver, July 16-21, pp. 263–267. BC, Canada (2006)
45. Gustafson, E.E., Kessel, W.C.: Fuzzy clustering with a fuzzy covariance matrix. IEEE CDC, San Diego, California, pp. 761–766. IEEE CDC, Los Alamitos (1979)
46. Hampel, F., Ronchetti, E., Rousseeuw, P., Stahel, W.: Robust Statistics: The Approach Based on Influence Functions. Wiley, Chichester (1986)
47. Hathaway, R.J.: Another interpretation of the EM algorithm for mixture distributions. Statistics & Probability Letters 4, 53–56 (1986)
48. Hathaway, R.J., Bezdek, J.C.: Switching regression models and fuzzy clustering. IEEE Trans. on Fuzzy Systems 1(3), 195–204 (1993)
49. Hathaway, R.J., Bezdek, J.C.: Fuzzy c-means clustering of incomplete data. IEEE Trans. on Systems, Man and Cybernetics Part B 31(5), 735–744 (2001)
50. Hathaway, R.J., Davenport, J.W., Bezdek, J.C.: Relational duals of the c-means clustering algorithms. Pattern Recognition 22(2), 205–212 (1989)
51. Herlocker, J.L., Konstan, J.A., Borchers, A., Riedl, J.: An algorithmic framework for performing collaborative filtering. In: Proc. of Conference on Research and Development in Information Retrieval (1999)
52. Hinton, G.E., Dayan, P., Revow, M.: Modeling the manifolds of images of handwritten digits. IEEE Trans. on Neural Networks 8(1), 65–74 (1997)
53. Holland, P.W., Welsch, R.E.: Robust regression using iteratively reweighted least-squares. Communications in Statistics A6(9), 813–827 (1977)
54. Honda, K., Ichihashi, H.: Linear fuzzy clustering techniques with missing values and their application to local principal component analysis. IEEE Transactions on Fuzzy Systems 12(2), 183–193 (2004)
55. Honda, K., Ichihashi, H.: Regularized linear fuzzy clustering and probabilistic PCA mixture models. IEEE Transactions on Fuzzy Systems 13(4), 508–516 (2005)
56. Honda, K., Ichihashi, H.: Component-wise robust linear fuzzy clustering for collaborative filtering. International Journal of Approximate Reasoning 37(2), 127–144 (2004)
57. Honda, K., Ichihashi, H.: Fuzzy local independent component analysis with external criteria and its application to knowledge discovery in databases. International Journal of Approximate Reasoning 42(3), 159–173 (2006)
58. Honda, K., Ichihashi, H., Masulli, F., Rovetta, S.: Linear fuzzy clustering with selection of variables using graded possibilistic approach. IEEE Transactions on Fuzzy Systems 15(5), 878–889 (2007)
59. Honda, K., Ichihashi, H., Ohue, M., Kitaguchi, K.: Extraction of local independent components using fuzzy clustering. In: Proc. of 6th Int. Conf. on Soft Computing, pp. 837–842 (2000)
60. Honda, K., Nakamura, Y., Ichihashi, H.: Simultaneous application of fuzzy clustering and quantification with incomplete categorical data. Journal of Advanced Computational Intelligence and Intelligent Informatics 8(4), 397–402 (2004)
61. Honda, K., Sugiura, N., Ichihashi, H., Araki, S.: Collaborative filtering using principal component analysis and fuzzy clustering. In: Zhong, N., Yao, Y., Ohsuga, S., Liu, J. (eds.) WI 2001. LNCS (LNAI), vol. 2198, pp. 394–402. Springer, Heidelberg (2001)

62. Honda, K., Togo, N., Fujii, T., Ichihashi, H.: Linear Fuzzy Clustering Based on Least Absolute Deviations. In: Proc. of 2002 IEEE Int Conf. on Fuzzy Syst., pp. 1444–1449 (2002)
63. Höppner, F., Klawonn, F., Kruse, R., Runkler, T.: Fuzzy Cluster Analysis. John Wiley & Sons, Chichester (1999)
64. Huber, P.J.: Robust Statistics. Wiley, Chichester (1981)
65. Huber, P.J.: Projection pursuit. The Annals of Statistics 13(2), 435–475 (1985)
66. Hyvärinen, A., Karhunen, J., Oja, E.: Independent Component Analysis. John Wiley & Sons, Ing., Chichester (2001)
67. Hyvärinen, A., Oja, E.: A fast fixed-point algorithm for independent component analysis. Neural Computation 9, 1483–1492 (1997)
68. Ichihashi, H., Honda, K., Tani, N.: Gaussian mixture PDF approximation and fuzzy c-means clustering with entropy regularization. In: Proc. of Fourth Asian Fuzzy Systems Symposium, vol. 1, pp. 217–221 (2000)
69. Ichihashi, H., Miyagishi, K., Honda, K.: Fuzzy c-means clustering with regularization by K-L information. Proc. of 10th IEEE International Conference on Fuzzy Systems 2, 924–927 (2001)
70. Inokuchi, R., Miyamoto, S.: LVQ Clustering and SOM Using a Kernel Function. In: Budapest, H.C.-R. (ed.) Proc. of FUZZ-IEEE 2004, Budapest, Hungary, July 25-29, 2004, pp. 1–4 (2004) (CD-ROM)
71. Inokuchi, R., Nakamura, T., Miyamoto, S.: Kernelized Cluster Validity Measures and Application to Evaluation of Different Clustering Algorithms. In: Proc. of 2006 IEEE International Conference on Fuzzy Systems, Sheraton Vancouver Wall Centre Hotel, Vancouver, BC, Canada, July 16-21, pp. 3660–3666 (2006)
72. Jain, A.K., Dubes, R.C.: Algorithms for Clustering Data. Prentice Hall, Englewood Cliffs, NJ (1988)
73. Jajuga, K.: L_1-Norm Based Fuzzy Clustering. Fuzzy Sets and Syst. 39, 43–50 (1991)
74. Joachims, T.: Making large-scale SVM learning practical. In: Scholkopf, B., et al. (eds.) Advances in Kernel Methods: Support Vector Learning, pp. 169–184. MIT Press, Cambridge (1999)
75. Jolliffe, I.T.: Discarding variables in a principal component analysis. I. Artificial data. Appl. Statist. 21, 160–173 (1972)
76. Kambhatla, N., Leen, T.K.: Dimension reduction by local principal component analysis. Neural Computation 9(7), 1493–1516 (1997)
77. Karhunen, J.: Neural approaches to independent component analysis and source separation. In: Proc. 4th European Symposium on Artificial Neural Networks (ESANN 1996), pp. 249–266 (1996)
78. Karhunen, J., Malaroiu, S., Ilmoniemi, M.: Local linear independent component analysis based on clustering. Int. Jour. of Neural Systems 10(6), 439–451 (1999)
79. Karhunen, J., Oja, E., Wang, L., Vigario, R., Joutsensalo, J.: A class of neural networks for independent component analysis. IEEE Trans. on Neural Networks 8, 486–504 (1997)
80. Kaufman, L., Rousseeuw, P.J.: Finding Groups in Data: An Introduction to Cluster Analysis. Wiley, New York (1990)
81. Kawaguchi, M.: Introduction to Multivariate Analysis I. Morikita Publishing Co. (1973) (in Japanese)
82. Kearns, M.J.: Thoughts on Hypothesis Boosting. ML class project (1988)
83. Kendall, M., Stuart, A., Ord, J.K.: The Advanced Theory of Statistics, 4th edn., vol. 3. Griffin, London (1983)

84. Kivinen, J., Warmuth, M.: Boosting as Entropy Projection. In: Proc. 12th Ann. Conf. Computational Learning Theory, pp. 134–144 (1999)
85. Kohonen, T.: Self-Organizing Maps, 2nd edn. Springer, Berlin (1997)
86. Konstan, J.A., Miller, B.N., Maltz, D., Herlocker, J.L., Gardon, L.R., Riedl, J.: Grouplens: Applying collaborative filtering to usenet news. Communications of the ACM 40(3), 77–87 (1997)
87. Krishnapuram, R., Keller, J.M.: A possibilistic approach to clustering. IEEE Trans. on Fuzzy Systems 1, 98–110 (1993)
88. Knuth, D.E.: Fundamental Algorithms, 2nd edn. Addison-Wesley, Reading (1973)
89. Lee, T.-W.: Independent Component Analysis - Theory and Applications. Kluwer, Dordrecht (1998)
90. Li, R.-P., Mukaidono, M.: A maximum entropy approach to fuzzy clustering. In: Proc. of the 4th IEEE Intern. Conf. on Fuzzy Systems (FUZZ-IEEE/IFES 1995), Yokohama, Japan, March 20-24, 1995, pp. 2227–2232 (1995)
91. Li, R.P., Mukaidono, M.: Gaussian clustering method based on maximum-fuzzy-entropy interpretation. Fuzzy Sets and Systems 102, 253–258 (1999)
92. Little, R.J., Rubin, D.B.: Statistical Analysis with Missing Data, 2nd edn. John Wiley and Sons, New York (2002)
93. Luo, F.-L., Unbehauen, R., Cichocki, A.: A minor component analysis algorithm. Neural Networks 10(2), 291–297 (1997)
94. Lusternik, L.A., Sobolev, V.J.: Elements of Functional Analysis. Hindustan Publishing, Delhi, India (1974)
95. MacQueen, J.B.: Some methods of classification and analysis of multivariate observations. In: Proc. of 5th Berkeley Symposium on Math. Stat. and Prob., pp. 281–297 (1967)
96. Masulli, F., Artuso, M., Bogus, P., Schenone, A.: Fuzzy clustering methods for the segmentation of multivariate medical images. In: Proc. of the 7th International Fuzzy Systems Association World Congress (IFSA 1997), Prague, Czech, June 25-30, 1997, pp. 123–128 (1997)
97. Masulli, F., Rovetta, S.: Soft transition from probabilistic to possibilistic fuzzy clustering. IEEE Transactions on Fuzzy Systems 14(4), 516–527 (2006)
98. McLachlan, G., Peel, D.: Finite Mixture Models. Wiley, New York (2000)
99. Miyamoto, S.: Fuzzy Sets in Information Retrieval and Cluster Analysis. Kluwer, Dordrecht (1990)
100. Miyamoto, S., Takata, O., Umayahara, K.: Handling missing values in fuzzy c-means. In: Proc. 3rd Asian Fuzzy Systems Symposium, pp. 139–142 (1998)
101. Miyamoto, S., Agusta, Y.: An Efficient Algorithm for l_1 Fuzzy c-Means and Its Termination. Control and Cybern. 24(4), 421–436 (1995)
102. Miyamoto, S., Mukaidono, M.: Fuzzy c-means as a regularization and maximum entropy approach. In: Proc. of the 7th International Fuzzy Systems Association World Congress (IFSA 1997), Prague, Czech, June 25-30, 1997, vol. II, pp. 86–92 (1997)
103. Miyamoto, S., Agusta, Y.: Algorithms for L_1 and L_p fuzzy c-means and their convergence. In: Hayashi, C., et al. (eds.) Data Science, Classification, and Related Methods, pp. 295–302. Springer, Tokyo (1997)
104. Miyamoto, S., Umayahara, K.: Fuzzy clustering by quadratic regularization. In: Proc. of FUZZ-IEEE 1998, Anchorage, Alaska, May 4–9, 1998, pp. 1394–1399 (1998)
105. Miyamoto, S.: Introduction to Cluster Analysis: Theory and Applications of Fuzzy Clustering, Morikita-Shuppan, Tokyo (1999) (in Japanese)

106. Miyamoto, S., Umayahara, K.: Methods in hard and fuzzy clustering. In: Liu, Z.Q., Miyamoto, S. (eds.) Soft Computing and Human Centered Machines, pp. 85–129. Springer, Tokyo (2000)
107. Miyamoto, S., Nakayama, Y.: Algorithms of hard c-means clustering using kernel functions in support vector machines. Journal of Advanced Computational Intelligence and Intelligent Informatics 7(1), 19–24 (2003)
108. Miyamoto, S., Suizu, D.: Fuzzy c-means clustering using kernel functions in support vector machines. Journal of Advanced Computational Intelligence and Intelligent Informatics 7(1), 25–30 (2003)
109. Miyamoto, S., Mizutani, K.: Fuzzy multiset space and c-means clustering using kernels with application to information retrieval. In: De Baets, B., Kaynak, O., Bilgiç, T. (eds.) IFSA 2003. LNCS (LNAI), vol. 2715, pp. 387–395. Springer, Heidelberg (2003)
110. Miyamoto, S., Suizu, D., Takata, O.: Methods of Fuzzy c-Means and Possibilistic Clustering Using a Quadratic Term. Scientiae Mathematicae Japonicae 60(2), 217–233 (2004)
111. Miyamoto, S., Kurosawa, N.: Controlling Cluster Volume Sizes in Fuzzy c-Means Clustering. In: Proc. of SCIS & ISIS 2004. Yokohama, Japan, September 21-24, 2004, pp. 1–4 (2004) (CD-ROM Proc. WP-3-1)
112. Miyamoto, S., Mizutani, K.: Fuzzy multiset model and methods of nonlinear document clustering for information retrieval. In: Torra, V., Narukawa, Y. (eds.) MDAI 2004. LNCS (LNAI), vol. 3131, pp. 273–283. Springer, Heidelberg (2004)
113. Miyamoto, S., Yasukochi, T., Inokuchi, R.: A Family of Fuzzy and Defuzzified c-Means Algorithms. In: Proc. of the International Conference on Computational Intelligence for Modelling, Control and Automation, Vienna, Austria, pp. 170–176 (2005)
114. Mizutani, K., Miyamoto, S.: Fuzzy Multiset Model for Information Retrieval and Clustering Using a Kernel Function. In: Zhong, N., Raś, Z.W., Tsumoto, S., Suzuki, E. (eds.) ISMIS 2003. LNCS (LNAI), vol. 2871, pp. 417–421. Springer, Heidelberg (2003)
115. Mizutani, K., Miyamoto, S.: Kernel-based Fuzzy Competitive Learning Clustering. In: Proc. of FUZZ-IEEE2005, The IEEE International Conference on Fuzzy Systems, Reno, Nevada, USA, May 22-25, 2005, pp. 636–639 (2005)
116. Mizutani, K., Miyamoto, S.: Possibilistic approach to kernel-based fuzzy c-means clustering with entropy regularization. In: Torra, V., Narukawa, Y., Miyamoto, S. (eds.) MDAI 2005. LNCS (LNAI), vol. 3558, pp. 144–155. Springer, Heidelberg (2005)
117. Mizutani, K., Inokuchi, R., Miyamoto, S.: Algorithms of nonlinear document clustering based on fuzzy set model. International Journal of Intelligent Systems (to appear)
118. Moerland, P.: A comparison of mixture models for density estimation. In: Proc. of 9th Int. Conf. Artificial Neural Networks (ICANN 1999), vol. 1, pp. 25–30 (1999)
119. Nock, R., Nielsen, F.: On weighting clustering. IEEE Transactions on Pattern Analysis and Machine Intelligence 28(8) (2006)
120. Oh, C.-H., Honda, K., Ichihashi, H.: Knowledge discovery from local principal components independent of arbitrary factors. In: Baba, N., Jain, L.C., Howlett, R.J. (eds.) Knowledge-Based Intelligent Information Engineering Systems & Allied Technologies, pp. 17–21. IOS Press, Amsterdam (2001)
121. Oja, E.: Principal components, minor components and liner neural networks. Neural Networks 5, 927–935 (1992)

122. Oja, E., Karhunen, J.: Signal separation by nonlinear Hebbian learning. In: Palaniswami, M., Attikiouzel, Y., Marks, R., Fogel, D., Fukuda, T. (eds.) Computational Intelligence - A Dynamic System Perspective, pp. 83–97. IEEE Press, New York (1995)
123. Okuno, T., Kume, H., Haga, T., Yoshizawa, T.: Multivariate analysis methods, JUSE Press (1971) (in Japanese)
124. Omachi, S., Sun, F., Aso, H.: A noise-adaptive discriminant function and its application to blurred machine-printed kanji recognition. IEEE Trans. on PAMI 22(3), 314–319 (2000)
125. Osiński, S., Weiss, D.: Conceptual clustering using Lingo algorithm: Evaluation on open directory project data. In: Proc. Int'l IIS: Intelligent Information Processing and Web Mining Conf, pp. 369–378. Springer, Heidelberg (2004)
126. Pal, N.K., Bezdek, J.C., Tsao, E.C.-K.: Generalized clustering networks and Kohonen's self-organizing scheme. IEEE Trans. on Neural Networks 4(4), 549–557 (1993)
127. Pal, N.R., Pal, K., Keller, J.M., Bezdek, J.C.: A possibilistic fuzzy c-means clustering algorithm. IEEE Transactions on Fuzzy Systems 13(4), 508–516 (2005)
128. Pearson, K.: On lines and planes of closest fit to systems of points in space. Philosophical Magazines and Journal of Science 2, 559–572 (1901)
129. Press, W.H., Teukolsky, S.A., Vetterling, W.T., Flannery, B.P.: Numerical Recipes in C, The Art of Scientific Computing, 2nd edn. Cambridge University Press, Cambridge (1999)
130. Quinlan, J.R.: Bagging, Boosting, and C4.5. In: Proc. 13th National Conference on Artificial Intelligence, AAAI/IAAI, Portland, Oregon, August 1996, vol. 1, pp. 725–730 (1996)
131. Redner, R.A., Walker, H.F.: Mixture densities, maximum likelihood and the EM algorithm. SIAM Review 26(2), 195–239 (1984)
132. Resnick, P., Iacovou, N., Suchak, M., Bergstrom, P., Riedl, J.: Grouplens: An open architecture for collaborative filtering of netnews. In: Proc. of ACM Conference on Computer-Supported Cooperative Work, pp. 175–186 (1994)
133. Rockafellar, R.T.: Convex Analysis. New Jersey. Princeton University Press, Princeton, New Jersey (1970)
134. Rose, K., Gurewitz, E., Fox, G.: A deterministic annealing approach to clustering. Pattern Recognition Letters 11, 589–594 (1990)
135. Rose, K.: Deterministic annealing for clustering, compression, classification, regression, and related optimization problems. Proc. of the IEEE 86(11), 2210–2239 (1998)
136. Roweis, S.: EM algorithms for PCA and SPCA. In: Jordan, M.I., Kearns, M.J., Solla, S.A. (eds.) Advances in Neural Information Processing Systems 10, pp. 626–632. MIT Press, Cambridge (1998)
137. Roubens, M.: Pattern classification problems and fuzzy sets. Fuzzy Sets and Syst. 1, 239–253 (1978)
138. Runkler, T.A., Bezdek, J.C.: Alternating cluster estimation: A new tool for clustering and function approximation. IEEE Trans. Fuzzy Syst. 7(4), 377–393 (1999)
139. Ruspini, E.H.: A new approach to clustering. Information and Control 15, 22–32 (1969)
140. Ruspini, E.H.: Numerical methods for fuzzy clustering. Information Sciences 2, 319–350 (1970)
141. Ruspini, E.H.: New experimental results in fuzzy clustering. Information Sciences 6, 273–284 (1973)

142. Salton, G., McGill, M.J.: Introduction to Modern Information Retrieval. McGraw-Hill, New York (1983)
143. Shardanand, U., Maes, P.: Social information filtering: Algorithms for automating word of mouth. In: Proc. of ACM Conference on Human Factors in Computing Systems, pp. 210–217 (1995)
144. Schapire, R.E., Singer, Y.: Improved Boosting Algorithms Using Confidence-Rated Predictions. In: Proc. 11th Int'l Conf. Computational Learning Theory, pp. 80–91 (1998)
145. Schölkopf, B., Warmuth, M.K. (eds.): COLT/Kernel 2003. LNCS (LNAI), vol. 2777. Springer, Heidelberg (2003)
146. Shibayama, T.: A PCA-like method for multivariate data with missing values. Japanese Journal of Educational Psychology (in Japanese) 40, 257–265 (1992)
147. Shibuya, K., Miyamoto, S., Takata, O., Umayahara, K.: Regularization and constraints in fuzzy c-means and possibilistic clustering. Journal of Japan Society for Fuzzy Theory and Systems (in Japanese) 13(6), 707–715 (2001)
148. Shum, H., Ikeuchi, K., Reddy, R.: Principal component analysis with missing data and its application to polyhedral object modeling. IEEE Trans. Pattern Analysis and Machine Intelligence 17, 854–867 (1995)
149. Sokal, R.R., Sneath, P.H.A.: Principles of Numerical Taxonomy. Freeman, New York (1963)
150. Späth, H.: Cluster Analysis Algorithms for Data Reduction and Classification of Objects, Ellis Horwood. Wiley, New York (1980)
151. Sun, F., Omachi, S., Aso, H.: Precise selection of candidates for hand written character recognition. IEICE Trans. Information and Systems E79-D(3), 510–515 (1996)
152. Swets, J.A.: Signal Detection Theory and ROC Analysis in Psychology and Diagnostics: collected papers. Lawrence Erlbaum, Mahwah (1996)
153. Tanaka, Y., Mori, Y.: Principal component analysis based on a subset of variables: Variable selection and sensitivity analysis. American Journal of Mathematics and Management Sciences 17(1&2), 61–89 (1997)
154. Tihonov, A.N., Arsenin, V.Y.: Solutions of Ill-Posed Problems. Wiley, New York (1977)
155. Timm, H., Kruse, R.: Fuzzy cluster analysis with missing values. In: Proc. 17th NAFIPS Int. conf., pp. 242–246 (1998)
156. Tipping, M.E., Bishop, C.M.: Mixtures of probabilistic principal component analysers. Neural Computation 11(2), 443–482 (1999)
157. Tsuchiya, T.: A quantification method for classification of variables. The Japanese Journal of Behaviormetrics (in Japanese) 22(2), 95–109 (1995)
158. Todd, M.J.: The Computation of Fixed Points and Applications. Lecture Notes in Economics and Mathematical Systems, vol. 124. Springer, Berlin (1976)
159. Troyanskaya, O., et al.: Missing value estimation methods for dna microarrays. Bioinformatics 17(6), 520–525 (2001)
160. Umayahara, K., Miyamoto, S.: Fuzzy clustering for detecting linear structures by fuzzification of dimensional coefficients. In: Proc. of 8th IFSA World Congress, vol. 2, pp. 752–756 (1999)
161. van Rijsbergen, C.J.: Information Retrieval, 2nd edn. Butterworth, London (1979)
162. Vapnik, V.N.: Estimation of Dependences Based on Empirical Data. Springer, Berlin (1982)
163. Vapnik, V.N.: Statistical Learning Theory. Wiley, New York (1998)
164. Vapnik, V.N.: The Nature of Statistical Learning Theory, 2nd edn. Springer, New York (2000)

165. Veenman, C.J., Reinders, M.J.T.: The nearest sub-class classifier: A compromise between the nearest mean and nearest neighbor classifier. IEEE Transactions on PAMI 27(9), 1417–1429 (2005)
166. http://vivisimo.com/
167. Whittle, P.: On principal components and least square methods of factor analysis. Skand. Akt. 36, 223–239 (1952)
168. Wiberg, T.: Computation of principal components when data are missing. In: Proc. 2nd Symposium on Computational Statistics, pp. 229–236 (1976)
169. Windham, M.P.: Numerical classification of proximity data with assignment measures. J. Classification 2, 157–172 (1985)
170. Wu, N.: The Maximum Entropy Method. Springer, Berlin (1997)
171. Xie, X.L., Beni, G.: A Validity measure for Fuzzy Clustering. IEEE Trans. on Pattern Analysis and Machine Intelligence 13(4), 841–846 (1991)
172. Yabuuchi, Y., Watada, J.: Fuzzy principal component analysis and its application. Biomedical Fuzzy and Human Sciences 3, 83–92 (1997)
173. Yager, R.R., Filev, D.P.: Approximate clustering via the mountain method. IEEE Trans. on Syst., Man, and Cybern. 24(8), 1279–1284 (1994)
174. Yanai, H.: Factor analysis with external criteria. Japanese Psychol. Res. 12(4), 143–153 (1970)
175. Zadeh, L.A.: Fuzzy sets. Information and Control 8, 338–353 (1965)
176. Zadeh, L.A.: Similarity relations and fuzzy orderings. Information Sciences 3, 177–200 (1971)
177. Zhang, B.: Generalized k-Harmonic Means. Technical Report TRHPL- 2000-137, Hewlett Packard Labs (2000)
178. Zhang, B., Hsu, M., Dayal, U.: K-Harmonic Means - Spatial Clustering Algorithm with Boosting. Temporal, Spatial, and Spatio- Temporal Data Mining, 31–45 (2000)
179. Zhang, B., Hsu, M., Dayal, U.: Harmonic Average Based Clustering Method and System. US Patent 6 584, 433 (2000)
180. Zhang, T., Ramakrishnan, R., Livny, M.: BIRCH: An efficient data clustering method for very large databases. Technical Report, Computer Sciences Dept., Univ. of Wisconsin-Madison (1995)
181. Zhou, X.-H., Obuchowski, N.A., McClish, D.K.: Statistical Methods in Diagnostic Medicine. John Wiley & Sons, New York (2002)
182. UCI repository of machine learning databases and domain theories, ftp://ftp.ics.uci.edu/pub/machine-learning-databases
183. VASpca (VAriable Selection in Principal Component Analysis), http://mo161.soci.ous.ac.jp/vaspca/indexE.html

Index

GPSR Compliance

The European Union's (EU) General Product Safety Regulation (GPSR) is a set of rules that requires consumer products to be safe and our obligations to ensure this.

If you have any concerns about our products, you can contact us on ProductSafety@springernature.com

In case Publisher is established outside the EU, the EU authorized representative is:

Springer Nature Customer Service Center GmbH
Europaplatz 3
69115 Heidelberg, Germany

Batch number: 10755292

Printed by Printforce, the Netherlands